Extreme Weather and Financial Markets

Founded in 1807, John Wiley & Sons is the oldest independent publishing company in the United States. With offices in North America, Europe, Australia and Asia, Wiley is globally committed to developing and marketing print and electronic products and services for our customers' professional and personal knowledge and understanding.

The Wiley Trading series features books by traders who have survived the market's ever changing temperament and have prospered—some by reinventing systems, others by getting back to basics. Whether a novice trader, professional or somewhere in-between, these books will provide the advice and strategies needed to prosper today and well into the future.

For a list of available titles, visit our web site at www.WileyFinance.com.

Extreme Weather and Financial Markets

Opportunities in Commodities and Futures

LAWRENCE J. OXLEY

John Wiley & Sons, Inc.

Published by John Wiley & Sons, Inc., Hoboken, New Jersey.
Published simultaneously in Canada.

Library of Congress Cataloging-in-Publication Data:

Oxley, Lawrence J., 1969–
Extreme weather and financial markets : opportunities in commodities and futures / Lawrence J. Oxley.
p. cm.—(Wiley trading ; 538)
Includes index.
ISBN 978-1-118-14721-4 (cloth); ISBN 978-1-118-20445-0 (ebk);
ISBN 978-1-118-20446-7 (ebk); ISBN 978-1-118-20447-4 (ebk)
1. Commodity futures. 2. Investment analysis. I. Title.
HG6046.O85 2012
332.64′5—dc23

2011041418

Printed in the United States of America

10 9 8 7 6 5 4 3 2 1

This book is dedicated to my three beautiful children, Brittany, Megan, and Jeffrey Oxley.

Contents

Preface

Extreme weather is hitting all regions of the globe with increasing severity. Despite the damage that can and will be caused from these extreme weather events, certain industries will nevertheless benefit and certain industries will be hurt. It is the purpose of this book to identify and evaluate the sectors, industries, companies, and more specifically the particular stocks, bonds, and futures that will be the winners and losers as extreme weather events continue to impact the Earth. Every investment idea in this book will work under the current, global climate condition. To the extent that these already existent extreme weather events get worse via global climate change, the more lucrative the investment ideas in this book become. The specialized definition of global warming as it relates to extreme weather investing is described here.

Definition: Global Warming

The rising temperature of both the air and the oceans as a result of a greenhouse effect due to excess pollution of carbon dioxide and other greenhouse gases, leading to global climate shocks and extreme weather events.

We have all been made very aware that the average global temperature will rise as a result of global warming. When things like gasoline, natural gas, and oil burn, they produce CO_2 (carbon dioxide), among other greenhouse gases. The CO_2 accumulates in the Earth's atmosphere, producing a greenhouse effect on the Earth, thus driving up the average temperature of both the air and the oceans over time.

However, that is only half the story. The burning of gasoline, natural gas, and oil produces not only CO_2 but also water. So we are steadily also increasing the quantity of water into the system that never existed before. In addition, as the average air temperature in the atmosphere rises, the more water the air is capable of holding. We will refer to this as *global watering*. This global watering, in combination with the increasing average temperature, global warming, will have the effect of causing global climate change. This could mean more rain, more snow, more ice, more droughts, and more severe weather such as hurricanes and tornadoes. It could even cause geographic shifts in weather patterns that are completely new to particular regions of the world. These effects are cumulative and will build over time and therefore do not represent a cycle where things return back to normal. The rapid growth of the emerging regions of the world, including China and India, will exacerbate these effects. It is the persistence of this pattern that provides the basis for this book.

The investment ideas in this book are geared for investors of all skill levels from the beginner right through to the professional investor. It is written in very straightforward and simple language, thus making the concepts and ideas very easy to understand.

The chapters are split out by the various types of extreme weather events to help the reader rapidly locate the applicable, relevant investment ideas during extreme weather events. The detailed index also provides assistance. Most importantly, the book is completely full of *action plan* tables that point out the specific company and commodity "biggest winners" and "biggest losers" resulting from extreme weather events occurring anywhere in the world!

The inclusion of real-life examples and specific investing rules for the extreme weather–based investor, as well as simplified tutorials on the basics of stock, bond, and futures market investing, all provide the reader with the necessary tools to make money on the extensive list of investment ideas discussed in this book.

Lawrence J. Oxley

Acknowledgments

Many people have played a part in the production of this book. In particular, my sincere thanks goes out to Lynda Oxley, Theresa Vitale, Sheila Doerr, and the numerous colleagues I have worked with both in my years as an engineer and in my years as an investor.

L.J.O.

CHAPTER 1

Commodities and Their Current Stories

As we sit here waiting for the extreme global climate events to begin to hit Planet Earth, we realize that they are already here. The following headlines of events have already occurred, along with thousands of others:

"Floods Swamp Eastern Australia"
"Droughts in Russia"
"Repeated Blizzards Cover the Northeastern Section of the United States"
"Hurricane Devastates Southern States"
"Vancouver Breaks Record for Coldest Freeze"
"Drier-than-Normal Conditions Lead to Dozens of Fires Destroying Timberlands"
"Heavy Rains Linked to Humans in Recent Study"
"Global Change Leads to Excessive Rain, Snow, and Flooding"
"Mudslides and Floods Destroy Homes"
"Sydney Sets Heat Wave Record"

There are many cynics in the world who do not believe that global climate change is occurring. The beauty of it, from an investor's standpoint, is that it doesn't even matter because, as shown in the preceding news events, we already have ample extreme weather events that will impact the stock market (and the bond and futures markets as well). To the extent that the environmental scientists are correct and the effects of global climate change do in fact get more frequent and powerful, the number of weather-based investment opportunities will increase significantly. Every investment idea in this book will work under today's global climate condition and will get even more lucrative if global climate change continues.

So what is the tie that binds extreme weather and financial markets? The tie that binds these two things together is commodity supply shocks. The extreme weather-based specialized definition of supply shock is one of the most critical things to understand in this entire book. It is the link between extreme weather events and the stock market. Let me explain.

Definition: Supply Shock

The temporary and sometimes permanent elimination of the supply availability of a commodity resulting primarily from a global climate shock or extreme weather event. The resultant spike in the price of the affected commodity provides opportunity for investment in multiple financial markets including primarily the stock market, the bond market and the futures market. Equivalent price spike reactions can also occur as a result of political, organizational, and major geologic events such as earthquakes.

It is also critically important to understand the extreme weather–based specialized definition of a commodity.

Definition: Commodity

Bulk goods and basic materials including metals, grains, food, minerals and energy, all derived from natural resources, which may or may not trade in the futures market. The price is subject to the forces of supply and demand. Price is particularly sensitive to supply shocks associated with extreme weather events.

It is also critically important to understand the newly created definition of *shock value.*

Definition: Shock Value

The amount of time a supply shock persists. The higher the number of days, the greater is our window of time for entry into an investment. The higher the number of days, the greater our potential holding period of the investment.

As shown in the definitions, a sudden change in the availability or "supply" of a commodity causes the price of that commodity to change. More specifically, if the supply of a material suddenly vanishes (due to an extreme weather event such as a flood in eastern Australia, which is a recent actual event that blocked the availability of coal in the region), then the price will go higher because everyone will be fighting to get the remaining supply and will be willing to pay more for it. The big winners in this case, of course, are the remaining producers of the commodity in short supply that are not involved in the devastated region. This is because they now get to sell their product at a higher price. Higher prices mean higher profits, and higher profits mean higher stock prices.

Now that we know that commodities are such a critical part of extreme weather–based investing, it's important for us to increase our general understanding of the various types of commodities and their relative attractiveness. The attractiveness of a commodity from an investor's standpoint is referring to the current supply/demand situation for that particular commodity. Examples always help to clarify what is meant by the attractiveness of a commodity. In the first example, we talk about the impact sugar prices have on the chocolate maker Hershey; and in the second example, we talk about a retail clothing store.

If the Hershey Company desperately needs sugar to make candy bars but sugar has suddenly become unavailable because of recent heavy flooding, the people at Hershey start to get very nervous because without the sugar they will not be selling any candy bars. The *fear* that Hershey feels causes them to increase the price they are willing to pay for sugar because if they do not get the sugar, it will be very costly to them when they stop selling their product. This is bad for Hershey because they are buying sugar, but it is a *jackpot* for the sugar makers because the price of sugar is going up. So, a sugar shortage results in decreased profits for Hershey and increased profits for sugar makers.

Another classic example is retail clothing excess inventory after the holidays have come to an end. In this case, the store desperately wants to get rid of the excess inventory to make room for the new season and the new fashions. Meanwhile, the holidays are already over and no one is buying. This combination of too much supply and a shrinking demand is a *big problem* for the retailer because the price of the clothing will need to go much, much lower to make the product sell. These two examples can be captured in the generalized supply/demand table shown in Figure 1.1. As shown in the table, when supplies are limited and demand is strong, this is the best possible combination for any commodity and is considered a jackpot, thus allowing the profits of the commodity producer to rise.

Similarly, when supplies are high and demand is relatively weak, this is a big problem for any commodity because the price of that commodity is going to fall hard, thus causing the profits for the makers of that commodity to drop. The remaining two possible combinations are considered *neutral* because supply and demand are both going in the same direction.

	Lots of Supply Coming	Very Little Supply Coming
Demand Growing	Neutral	Jackpot
Demand Shrinking	Big Problem	Neutral

FIGURE 1.1 Commodity Categories

For example, it is acceptable if supplies are growing as long as demand is growing, too; and it is acceptable if demand is shrinking as long as the supply side is staying stable to declining.

The exciting part about commodities is that the vast majority of them fall into either the Jackpot or Neutral category and very few fall into the Big Problem category. We will get into more detail shortly, but for now we will simply list the major commodities and categorize them into the quadrants shown in the table.

Jackpot

- Copper
- Corn
- Wheat
- Soybeans
- Orange juice
- Fertilizers
- Metallurgical coal
- Iron ore
- Oil
- Sugar
- Cocoa beans
- Cotton
- Coffee
- Silver
- Gold
- Platinum
- Palladium
- Rare earths
- Potash

Neutral

- Zinc
- Aluminum
- Steel
- Nickel
- Lead
- Thermal coal
- Soda ash

Big Problem

- Natural gas

Limited Opportunity

- Lean hogs
- Live cattle

Granted, commodities are cyclical and their status changes over time, but their current cycle is very sticky, meaning that global demand for commodities in general is growing (over the long haul) and simultaneously it is getting harder and harder to find a good, quality supply of these materials. This combination means the outlook for commodities over the long haul is generally good. Now, what happens when a particular commodity over the next 10 years happens to move from the Jackpot category to the Big Problem category? Does that destroy our investment opportunities we talk about in this book? No, absolutely not. Regardless of where a commodity falls in the four categories, it will still respond quite favorably (i.e., price of the commodity will go up and therefore the profits for the commodity maker will go up, and hence the stock price of the commodity maker will go up) in the event of a supply shock associated with global climate change. Nevertheless, I include a discussion on the supply/demand status of each commodity to help the reader understand these materials even better. In addition, even though supply shocks from global climate change will help any commodity, it is particularly juicy from an investing standpoint if it falls within the Jackpot category.

JACKPOT COMMODITIES

Now we will go into some more detail on the story behind each of the major commodities, starting with the Jackpot commodities.

Copper

Copper has among the healthiest fundamental outlooks for commodity materials. The supply/demand balance for copper remains tight despite the fact that the United States residential and nonresidential construction markets remain soft but at least are finally nearing the bottom.

Figure 1.2 shows the end-market usage for copper. As the demand for copper in China continues to grow and as construction demand in the United States rebounds, the demand side of copper will become increasingly attractive.

Despite very strong demand in China for copper, copper mine concentrations are not in China, thus greatly helping the supply-side story. Remember, there are a lot of people in China who need jobs. If copper mining were geologically abundant in China, the decision to add jobs by adding mining capacity would occur, but instead this is not a problem for this commodity. Another supply-side dynamic that is favorable to copper is the decreasing quality of the copper ore coming out of the mines. So, in many cases around the world, despite increasing rock movement in the mines, less copper is produced from the prior year, thus continuing to exacerbate an already tight supply/demand balance, thus putting copper squarely into the Jackpot category.

The global climate shock–type investment opportunities for copper reside in the stock market, the bond market, and the futures market, and to a much lesser extent in the currency market, as we will cover in more detail later.

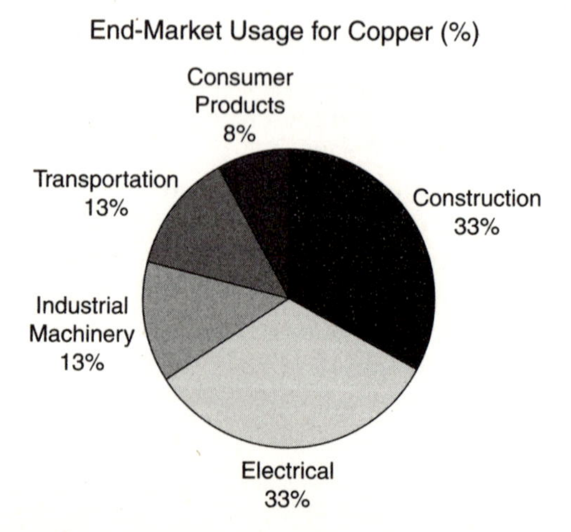

FIGURE 1.2 End-Market Usage for Copper
Source: Freeport McMoRan public filings, 2010.

Corn

Corn, like copper, has both a favorable demand-side story and a favorable supply-side story, thus again putting it squarely into the Jackpot category. On the demand side, global population growth guarantees the need for more food (see Figure 1.3 for the demand uses for corn).

Food is a nondiscretionary purchase. However, the increasing demand for food gets turbocharged when looking at what's happening in emerging regions such as China. As the population in China gets wealthier, they tend to migrate from eating strictly grain to instead incorporating beef, chicken, and pork. It takes approximately eight pounds of grain to make one pound of beef. It is this shifting dynamic in food consumption that provides the exponential upward slope in the demand curve. In addition, within the United States, the decision has already been made to add up to 15 percent ethanol to the gasoline pool. Ethanol is derived from corn. The need to convert corn into ethanol to satisfy demand in the gasoline market puts even more demand pressure onto corn. So the demand side of the corn analysis is obviously a home run.

On the supply side, the story also looks good. Within the United States, the opportunity to increase the available acres of farmland for growing corn in the United States is very limited. The United States represents a whopping 41 percent of global corn production. Even on a global scale, Brazil is among the few places in the world where incremental farmland acres are available. The tight supply and strongly growing demand for corn makes corn a very attractive commodity in general, but importantly it also meets the key criteria for making it susceptible to global climate shocks. Specifically, it is produced in highly concentrated regions of the world (i.e.,

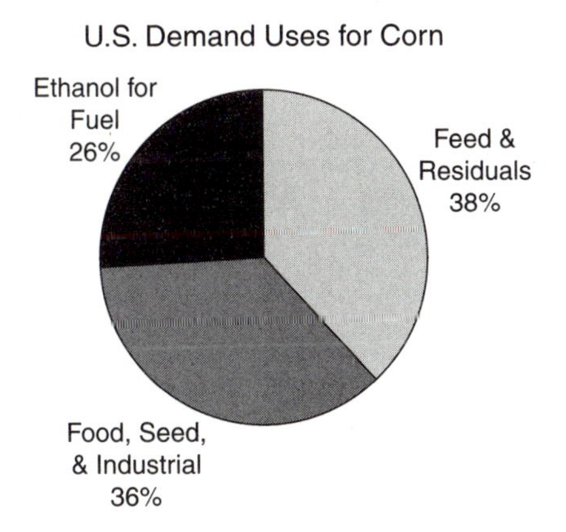

FIGURE 1.3 U.S. Demand Uses for Corn
Source: USDA, 2009.

the United States produces 41 percent of the global total) and therefore if the Corn Belt within the United States sees widespread drought or flooding conditions, then the price of corn will increase significantly, thus providing us plenty of investment opportunities.

As will be shown, the primary weather-based investment opportunity in corn lies in the futures market. There are also indirect investment opportunities related to corn in the stock market and corporate bond market.

Wheat

The story behind wheat is similar to that of corn in terms of the nondiscretionary demand for food globally. In fact, it is second only to rice as the main human food crop and still ahead of corn. As global population grows, so, too, does the demand for wheat consumption (see Figure 1.4 for the domestic demand uses for wheat).

On the supply side, wheat is more diversified globally in production as compared with corn. Nevertheless, the geographic concentration of production is high enough to still allow the price of wheat to spike in the event of a global climate shock in a particular wheat-producing region of the world. The classic example was the severe drought in Russia in the summer of 2010. Although Russia represents only 9 percent of the global production of wheat, it was still a large enough quantity to drive up wheat prices. The bottom line for wheat is that the combination of a healthy demand side along with limited acreage on the supply side places wheat into the Jackpot category.

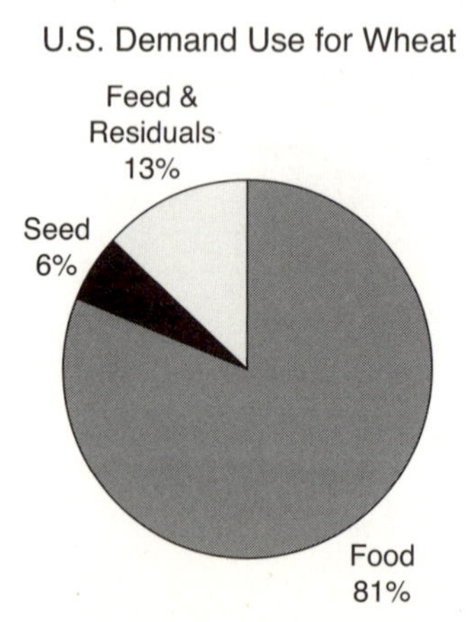

FIGURE 1.4 U.S. Demand Use for Wheat
Source: USDA, 2010.

The primary weather-based investment opportunity in wheat lies in the futures market.

Soybeans

The story behind soybeans is similar to that of corn in terms of the nondiscretionary demand for food globally. Therefore, there is generally a positive correlation between the demand for soybeans globally and global population. The link to population growth enables a healthy demand-side outlook for soybeans. However, the link to population growth is actually better than it appears. As we talked about in the corn section, the most exciting demand driver for soybeans is related to the migration of the daily diet of people in the emerging markets, such as China, toward higher protein intake in the form of beef, chicken, and pork as opposed to the historically traditional grain-based diets. As we talked about in the corn section, eight pounds of grains such as soybeans are required to produce only one pound of beef. This is a very powerful positive on the demand side for soybeans (see Figure 1.5 for the demand uses for soybeans).

On the supply side, soybeans have an exciting story, too. Their global, geographic concentration of production is high, meaning there are only a few geographies in the world where soybeans are made as well as a limited quantity of acres within each geography. This is very favorable because it lends itself to global climate shock–type investing. In other words, when a region of the world that produces soybeans is racked with severe drought or severe flooding, the price of soybeans goes up and

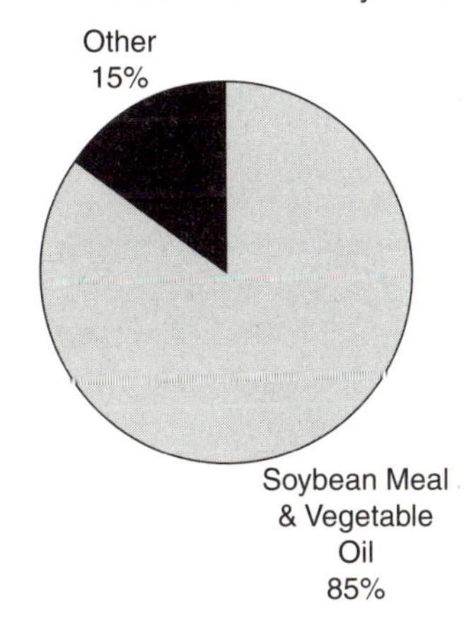

FIGURE 1.5 U.S. Demand Uses for Soybeans
Source: USDA, 2010.

therefore represents an opportunity for us (we will cover this in more detail later).

Another interesting dynamic that occurs on the supply side is farm acreage rotation between corn and soybeans, for example. When it comes time for springtime planting, the farmer has some choices to make. Using this example, the farmer must choose between corn and soybeans to plant on his available acreage. The decision on which one to plant depends on multiple factors. One key factor is the price of corn versus the price of soybeans. If the ratio of the price of soybeans to corn is very low, then it makes sense for the farmer to plant corn because he will maximize his profits by selling more corn instead of the relatively low-priced soybeans. In addition to the economic decision behind planting corn versus soybeans, the farmer also has other considerations, including, for example, the desire to rotate the crops to help get rid of certain pests in the soil that may attack soybeans more than corn. He also may rotate the crops to optimize the nutrients in the soil. Corn, for example, absorbs more nitrogen fertilizer than does soybeans partly because of the ability of soybeans to receive some of its nitrogen requirements directly from the nitrogen in the air. By rotating the crops the farmer optimizes his economics as well as technical aspects of the soil. However, by rotating the crops, the farmer impacts the supply availability of soybeans to the rest of the world. If, for example, in one particular spring, it makes sense for farmers to make more corn than soybeans, then soybeans may be in a shortage at harvest time. The bottom line for soybeans is that the existing demand- and supply-side stories puts soybeans squarely into the Jackpot category in Figure 1.1.

As we will cover in more detail later, the primary weather-based, direct investment opportunity in soybeans rests in the futures market (as opposed to the stock, bond, or currency markets).

Orange Juice

The orange juice story is generally quite healthy and we are not just talking health benefits here. On the demand side globally, consumption of oranges and orange juice is growing in alignment with population growth but also with the general interest in the health benefits of this fruit.

However, it's the supply side that makes the orange juice story particularly attractive. Florida, California, and Brazil combined produce a very large percentage of the oranges in the world. One might ask why there is such a limited geography in the production of oranges. A big part of the reason rests with the stringent growing conditions that must be present to grow oranges successfully. Oranges do best when grown in climates that have air temperatures in the 60 to 85 degree Fahrenheit range, thus limiting possible production locations globally. Interestingly, oranges are particularly sensitive to frost conditions. Under frost conditions, the farmer often has to resort to using portable heater pots in proximity to the orange crop and also, ironically, to spraying the crop with water, which keeps the temperature slightly above freezing.

The healthy demand and very limited supply-side stories for orange juice place this commodity into the Jackpot category.

As we will see, the extreme weather–based investment opportunity for orange juice rests within the futures market.

Fertilizers

Recall from the earlier discussion on corn that the amount of acreage available for the production of corn is quite limited. So what is a farmer to do if he wants to increase the amount of corn he produces each year? The farmer has no choice but to increase the amount of bushels per existing acre of farmland with the use of fertilizer. So, in other words, the bullish demand outlook for corn cascades down into the fertilizer space. All three of the major fertilizer types, including potash, phosphates, and nitrogen-based fertilizers, are in high demand because of the farmer's desire to increase his crop yields (bushels per acre of corn) in order to keep up with the ever-growing demand for corn.

What about the supply side of fertilizers? The short answer is that for all three primary types of fertilizer there is very limited supply coming on stream over the next few years. Without getting into too much detail, each of the three fertilizer types has its own particular reason why supply is limited. Briefly, for the mineral potash, the supply is limited because of the very limited number of global players in possession of the high-quality potash rock, the best of which is located in Saskatchewan, Canada (see separate section on potash rock).

As far as the phosphate fertilizers are concerned, the same basic story holds, with a very limited number of global players in control of the highest-quality phosphate rock.

Within the nitrogen fertilizer space, the capacity is limited by access to cheap natural gas, which is the primary raw material needed to manufacture nitrogen-based fertilizers. This limit is gradually going away, however, with the relatively recent discovery of excess natural gas supplies.

Another factor that it is important, with respect to limiting the available global fertilizer capacity, is the massive amount of money and time required to build additional mining and production capacity.

The bottom line for fertilizer is that the strong demand and the short supply put fertilizers into the Jackpot category.

The opportunities for investing in the fertilizer space are in the stock market and the corporate bond market, as we will see later in the book.

Metallurgical "Met" Coal (aka Coking Coal)

Metallurgical coal is another great story. Most coal in the world is used in the production of electricity. Metallurgical coal is special coal in the sense that it is better in terms of carbon content, lower impurities, and lower volatile material. This very high quality coal is not used to make electricity. It is mostly used in the production of steel, as shown in Figure 1.6.

Technically speaking, this coal must be specially prepared before it is used in the production of steel. For those interested in a bit of technology, this coal must first be heated in an inert atmosphere (inert means basically no oxygen allowed when baking this coal because the oxygen would react with the coal and make our coal turn into a gas and disappear from the baking oven—not exactly the best way to sell coal). After baking, we get a more stable and pure form of coal now known as "coke" at this stage of the production. This coke is used in steel-making blast furnaces. The coke reacts with iron ore in the blast furnace, producing iron metal, which is later converted into steel (steel is nothing more than iron plus a little carbon addition). So, technology aside, the key point here is that as the demand for steel grows globally, so, too, does the demand for metallurgical coal. Most groups around the world see global steel production growing at about 4 to 6 percent per year. This translates into similar levels of growth for "met" coal, thus making the demand side of the equation for "met" coal very attractive.

On the supply side, there are only a limited number of geographies that contain this high-grade coking coal, including eastern Australia, western Canada, and the eastern United States. Other regions of the world have the

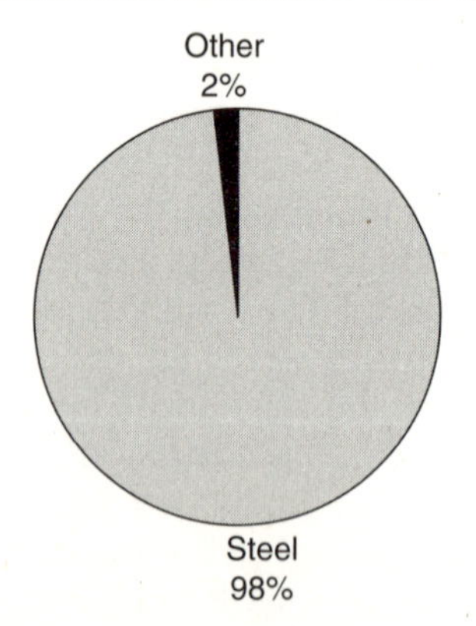

FIGURE 1.6 Global Market Demand for "Met" Coal
Source: Teck Resources public filing, 2009.

material but the quality is not as good, thus helping to limit the high-quality supply.

With demand strong and high-quality supply limited, this points to solid fundamentals for met coal and also puts met coal into the Jackpot category.

As we will see later in the book, weather-based investment opportunities in the met coal sector include the stock market and the corporate bond market. Opportunities to invest in met coal in the futures market and the foreign exchange market are essentially nonexistent.

Iron Ore

The story of iron ore is nearly identical to the story of coking coal. The three primary raw materials used to produce steel from scratch in a blast furnace (i.e., as opposed to simply remelting scrap steel in something called a "mini-mill") are iron ore, coke (aka processed met coal), and limestone. The iron ore is the key source of iron metal in the production of steel. Similar to metallurgical coal, the greatest demand driver for iron ore is also steel, as shown in Figure 1.7.

So, similar to metallurgical coal, as the global production of steel grows, the demand for iron ore grows correspondingly. Looking forward, global steel production will continue to grow at the rate of 4 to 6 percent per year, according to most public estimates. This is very good news for producers of iron ore.

On the supply side, it is difficult to find very high quality iron ore. In fact, there are only a few locations in the world with very high quality iron

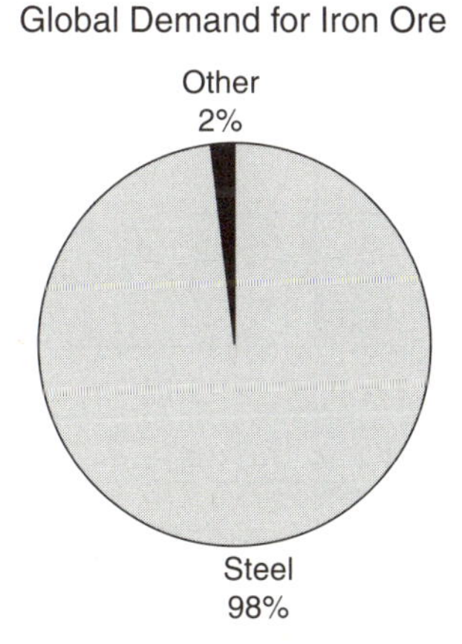

FIGURE 1.7 Global Demand for Iron Ore
Source: USGS, 2010.

ore, including Brazil and western Australia, for example. China has lots of iron ore but of the lower-quality type, hence the need to import this material into China. Thus, the supply side of the iron ore analysis also is generally quite favorable. New capacity in the western Australian region is a couple of years out into the future still.

The strong demand for iron ore in combination with limited supply makes the fundamentals of this material quite attractive and, in fact, quite similar to the metallurgical coal story, thus again putting this material into the Jackpot category.

As will be shown, the extreme weather–based investment opportunities in the iron ore space reside in the stock market and the corporate bond market. Opportunities to invest in iron ore within the futures market or within the foreign exchange market are largely nonexistent.

Oil

The global demand for oil—and every other energy source, for that matter—is projected to grow, according to the 2010 Energy Information Administration's (EIA) world energy report. Oil demand in particular is expected to grow by 1.4 percent per year globally. See Figure 1.8 to compare the demand growth of oil with other fuel types.

The positive demand growth story is also supported by the global emergence from the recession trough, which occurred in early 2009.

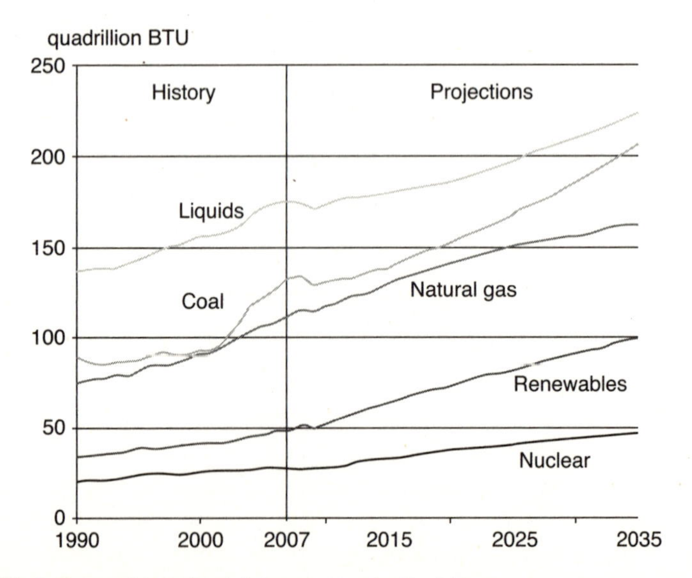

FIGURE 1.8 World Marketed Energy Use by Fuel Type
Source: EIA public filing, 2010.

On the supply side, it is getting harder and harder to find easy-access oil. The oil that we do find is either more costly and/or dirtier than the oil of the past, and it is generally found in deeper and deeper locations with higher and higher levels of sulfur content. In addition, on the supply side, the global oil market is plagued with politically driven global supply shocks.

Once again, the combination of growing global demand and tight supply leads to strong fundamentals for the oil commodity, thus again putting it into the Jackpot category.

Investment opportunities within the oil commodity reside in the futures market, the stock market, and the bond market.

Sugar

As emerging regions of the world continue to migrate toward a Western diet filled with processed food, soft drinks, and other confectionary delights, the global demand for sugar will continue to grow. Most foreign and domestic government sources point to global growth in the 2 percent per year range.

On the supply side of the sugar analysis, we see that sugar meets the extreme weather–based investing criteria of global, geographic concentration. In other words, a few select regions of the world hold a disproportionately high level of global sugar production. This is important because in the event of a global climate shock in the key sugar-producing regions of the world, we will see sugar prices climb rapidly.

The growing demand and favorable supply-side situation for sugar places sugar into the Jackpot category in Figure 1.1.

The weather-based investment opportunities in the sugar commodity lie mostly within the futures market. There are also a few select opportunities within the stock market, as will be explained later.

Cocoa

Cocoa is the key ingredient in the production of chocolate. The positive, global demand growth for chocolate is the result of population growth as well as the increased per-capita consumption of chocolate in emerging economies as the average income in these regions rises. “Emerging” regions of the world refer to places such as China, Russia, India, Brazil, and some areas of eastern Europe. Currently, mature regions such as Switzerland, Europe, and North America have the highest per-capita consumption of chocolate.

On the supply side, commodities do not get much more attractive than cocoa in terms of its very high geographic concentration. As will be discussed in much more detail later, western Africa dominates the production of the cocoa bean. There was a time when the Brazilian region was a much larger part of the global production of the cocoa bean. However, the crop disease known as Witch's Broom significantly lowered the crop yield in this area of the world. For this reason, and also because of cheaper labor costs, the western African region dominates cocoa bean production today.

The growing global demand and the favorable supply-side dynamic within the cocoa bean commodity places cocoa into the Jackpot category.

Weather-based investment opportunities in the cocoa bean reside predominantly within the futures market. There are also a few opportunities within the stock market as will be discussed later.

Cotton

Although most people are most familiar with cotton as a fiber used in the production of apparel, its end-use market split is actually a bit broader. The pie chart in Figure 1.9 shows the primary end-use markets for cotton.

The key demand driver for cotton is China. China drives demand for many commodities given its very high gross domestic product (GDP) in combination with its very low per-capita consumption of virtually every material. They consume 40 percent of the global demand for cotton. Despite the positive demand growth for cotton, there is some substitution risk from man-made fibers, including polyesters, as well as from other natural fibers, including wood fibers. Nevertheless, the global demand remains positive.

On the supply side, cotton has a fairly impressive level of geographic concentration of production. This is a prerequisite for effective global climate shock–type investing, as discussed repeatedly. Among the leaders in

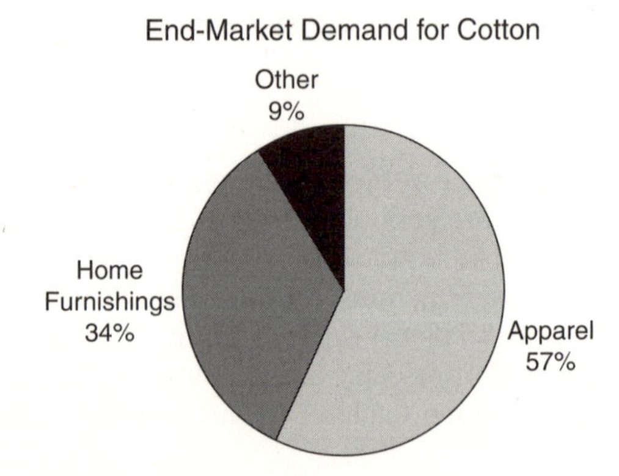

FIGURE 1.9 End-Market Demand for Cotton
Source: Cotton.org, 2010.

the production of cotton is China, with 29 percent of global cotton production (more on this later). The fact that China consumes more than it makes translates into healthy imports into China for this commodity.

The combination of globally healthy demand and a geographically concentrated supply puts cotton into the Jackpot category.

Weather-based investment opportunities with cotton reside predominantly within the futures market. Opportunities within the stock market, the corporate bond market, and the foreign exchange market are quite limited.

Coffee

Globally, the consumption of coffee is quite mature. The compounded annual growth rate shown in Figure 1.10 is 1.5 percent. Population growth and higher incomes contribute to demand growth but the mature regions already have fairly saturated per-capita consumption rates. Nevertheless, the demand curve remains slightly positive on a long-term best-fit basis, as shown in Figure 1.10.

On the supply side, coffee has the desirable geographic concentration in production, which lends itself to global climate shock–type investing. We will cover the supply side in more detail later.

The combination of a rugged but mature demand curve and the geographically concentrated supply side puts coffee into the Jackpot category.

Weather-based investment opportunities in coffee reside exclusively in the futures market. More on this later.

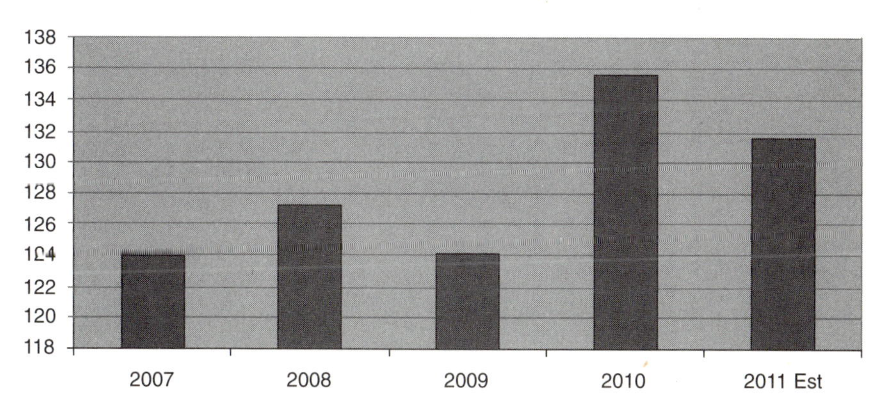

FIGURE 1.10 Global Coffee Consumption
Source: USDA, 2010.

Silver

The key end markets for silver are shown in Figure 1.11. As shown, the end-market demand is quite diversified and, in fact, mirrors GDP more so than it mirrors jewelry consumption, as one might expect. Aside from the secular decline occurring in digital conversion in the photography end market, the demand is growing overall, particularly after recovering from the declines in the global recession.

On the supply side, silver is produced as a by-product of lead/zinc/gold production, representing 70 percent of total mined product, and the balance comes from primary or "on purpose" silver mining production. If we include old silver scrap in the supply mix, then scrap holds 20 percent of global supply in 2010. Silver has a fairly concentrated geographic mix in terms of production quantity, thus making it fairly attractive for its extreme weather–based investment opportunities.

Given the very strong global demand and the fairly concentrated global geographic supply, silver gets placed into the Jackpot commodity category.

Extreme weather–based investment opportunities lay in the futures market predominantly with some additional opportunities in the stock market as well. More on this later.

Gold

Unlike silver, jewelry is the predominant demand driver for gold, as shown in Figure 1.12. Jewelry drives 69 percent of demand for gold.

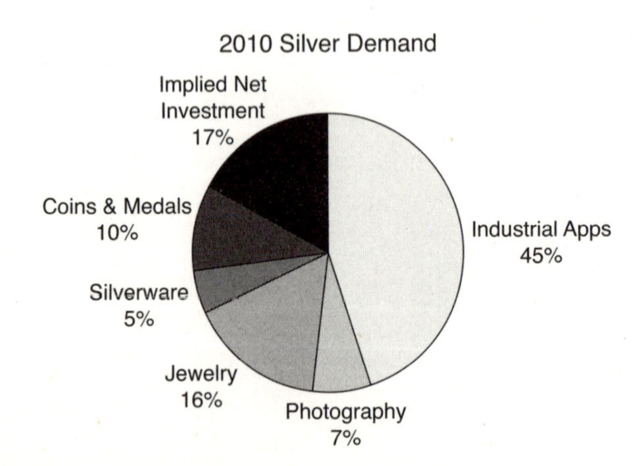

FIGURE 1.11 2010 Silver Demand
Source: Silverinstitute.org, 2010.

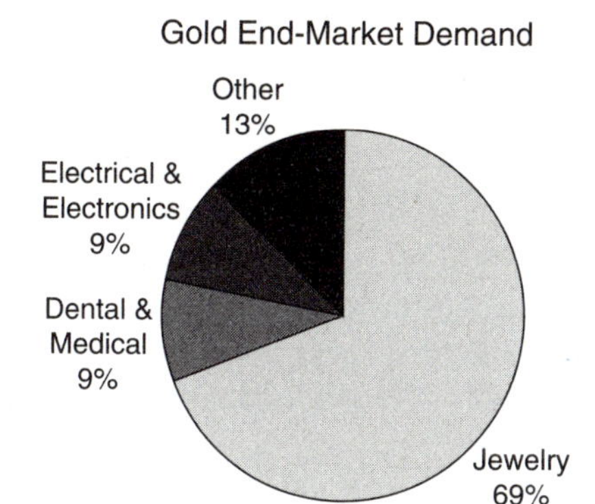

FIGURE 1.12 Gold End-Market Demand
Source: USGS, 2010.

The improving economics for individuals in emerging nations, including India and China, are helping drive the increase in demand for jewelry. Gold gets the double positive whammy in that not only is the functional demand for gold growing but gold also acts as a safe economic haven commodity. The countercyclical nature of gold allows the demand for gold to increase when the economy seems shaky and also when the value of the U.S. dollar gets weaker. Evidence of the strength in the gold commodity is glaringly obvious when looking at the price of gold over the past decade. Relative to the price in other commodities, the price of gold did not decline nearly as much during the global recession beginning in the fourth quarter of 2008. In fact, the price of gold is up 360 percent since 2002.

On the supply side, there are also constraints in terms of locating high-quality, high-yielding mining assets. The combination of the positive demand-side factors with the fairly tight supply side puts gold into the Jackpot category.

Interestingly, however, when evaluating gold from the standpoint of an extreme weather–based investment, it becomes much more challenging. Yes, it meets criterion number one of a favorable supply/demand outlook; however, it is challenged in terms of the second requirement of geographic concentration. Global gold production split out geographically is shown in Table 1.1.

The challenge for gold is that despite having 9 percent of its global production in the United States, the mining assets within the United States are geographically diverse. Therefore, an extreme weather event would have to hit a very large portion of the United States simultaneously. This makes gold much less attractive than corn, for example, given that 41 percent of global corn production comes from the United States, and even within the

TABLE 1.1 World Gold Production

World Gold Production	Percent Split
China	14%
Australia	10%
United States	9%
Russia	8%
South Africa	8%
Peru	7%
Indonesia	5%
Canada	4%
Uzbekistan	4%
Brazil	3%
Mexico	2%
Papua New Guinea	2%
Chile	2%
Other	21%
Total	**100%**

Source: USGS, 2010.

United States, the Midwest Corn Belt drives the lion's share of that total, thus making corn a very good investment opportunity from the standpoint of extreme weather–based climate shocks. We therefore will not spend a great deal of time on gold for extreme weather–based investing. Remember again, the purpose of this book is to identify investment opportunities in the context of extreme weather–based investing. Therefore, although gold may be a great investment in general due to its "jackpot"-type characteristics, it is not terribly attractive from the standpoint of an extreme weather–based investor.

Platinum

Platinum is a member of the platinum group metals (PGMs). The PGM group includes platinum, palladium, rhodium, ruthenium, and iridium. Platinum has a very exciting story for the extreme weather–based investor. First, let's take a look at the demand side. Globally, the key demand drivers for platinum are outlined in Figure 1.13.

As shown, the key drivers for platinum are auto catalysts and jewelry. Auto catalysts are used in the vehicle emission system. Their job is to reduce noxious emissions into the atmosphere. The key demand driver for catalysts is the global auto build rate, which is growing even in the United States currently, but more so in China. The demand for jewelry is also growing globally, in line with rising income levels, particularly in the

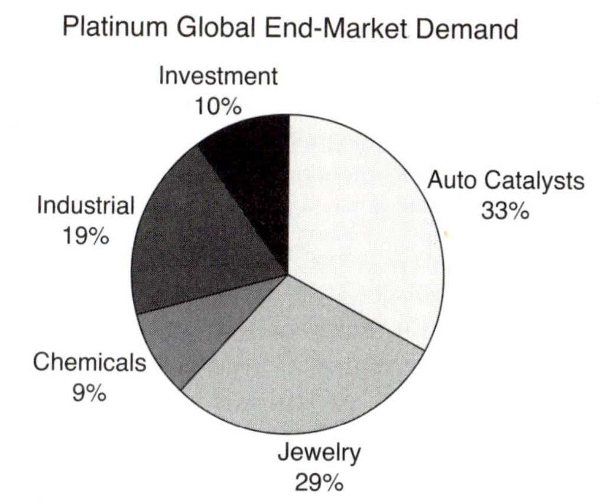

FIGURE 1.13 Platinum Global End-Market Demand
Source: Johnson Matthey public filings, 2010.

emerging regions of the world. On a geographic basis, the demand split is shown in Figure 1.14.

Two surprising pieces of information come out of the geographic demand split pie chart. First, the demand for catalysts is very high in Japan. This follows from the strong global position held by the Japanese automakers. Second, China also holds a very large portion of the global demand for platinum. This is impressive considering that the primary mode of transportation in China just two decades ago was the bicycle. Chinese demand for both cars and jewelry is growing strongly as income levels continue to

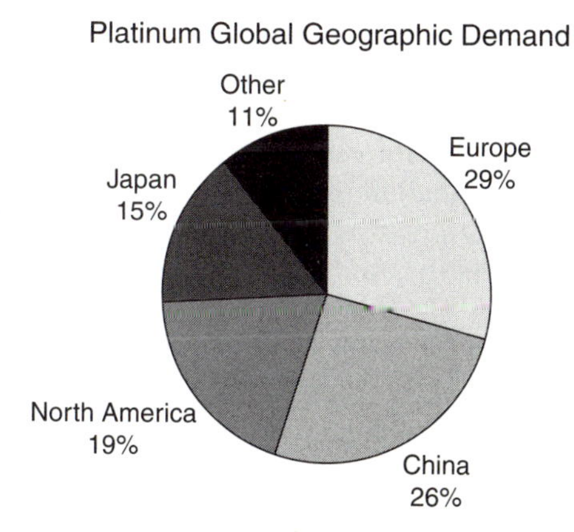

FIGURE 1.14 Platinum Global Geographic Demand
Source: Johnson Matthey public filings, 2010.

rise in this emerging region. Overall, the demand story for platinum is quite healthy, despite the demand-side interruption that occurred as a result of the monstrous Japanese earthquake in 2011.

On the supply side, things get even more exciting. Platinum has one of the most concentrated geographic sources of production of all commodities. This characteristic is critical for an extreme weather–based investor. As a reminder, we like very high geographic concentration in a commodity. In fact, the higher the level of geographic concentration, the better. This is true because in the event of a major global climate shock, which impacts a critical supply region for a particular commodity, the price of that commodity will rise. The higher the geographic concentration, the higher the price will rise. In addition, the longer the duration of the extreme weather event and the higher the impact on the supply availability of the commodity, the higher and longer the price will continue to rise.

Overall, given a healthy demand and a very consolidated supply, platinum gets placed into the Jackpot category.

We will cover platinum in much more detail in the chapter covering global climate shocks affecting mines. Extreme weather–based investment opportunities exist for platinum in the stock market, the futures market, and the exchange-traded fund (ETF) market. This commodity ranks among the highest in terms of its extreme weather–based investing attractiveness and opportunities.

Palladium

Palladium is also within the PGM series. Geologically, all of the PGMs are often found together. This is the reason that the producers of platinum are also the producers of palladium almost exclusively. Despite the similar geologic characteristics, palladium has a slightly different end-market demand split. Figure 1.15 shows the global end-market demand split for the metal palladium.

As shown, the demand for auto catalysts is the key driver for palladium. In fact, because palladium is cheaper than platinum on a per-ounce basis, palladium has taken market share away from platinum. Despite the shift in market share, the demand for both metals in the auto catalyst application is still growing.

By contrast, the jewelry end market favors platinum over palladium at this time. Outside of the differences in the auto and jewelry end markets, these two metals have similar demand drivers. The geographic end market demand for palladium is shown in Figure 1.16.

Similar to the global geographic demand split for platinum, aside from the developed regions of the world in North America and Europe, the other two key geographies are Japan and China. Japan is critical because of the

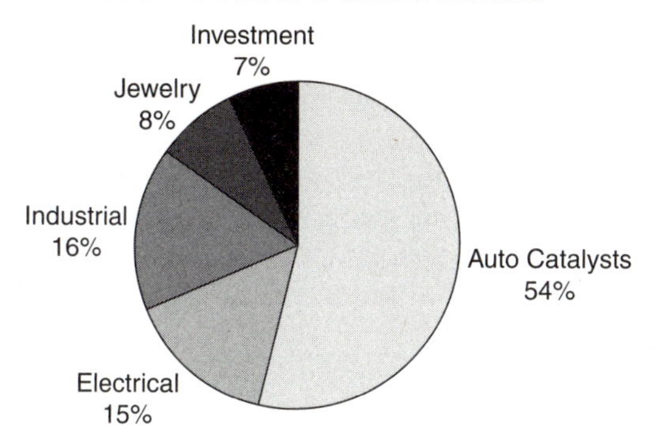

FIGURE 1.15 Palladium Global End-Market Demand
Source: Johnson Matthey public filings, 2010.

excellent global position held by the Japanese automakers. Palladium auto catalyst demand obviously correlates with the auto build rate. China is critical as the strongest global demand driver not only for palladium but for the vast majority of commodities in general. Interestingly, even within the developed regions of the world, palladium demand growth is solid not only because of the car build rate but also because of the increasingly challenging auto emission standards and the consequent need for auto catalysts. So, in general, the demand story is very good for palladium.

The supply side for palladium, like the supply side for platinum, is downright exciting for the extreme weather–based investor. It possesses the very desirable, highly concentrated global production characteristics

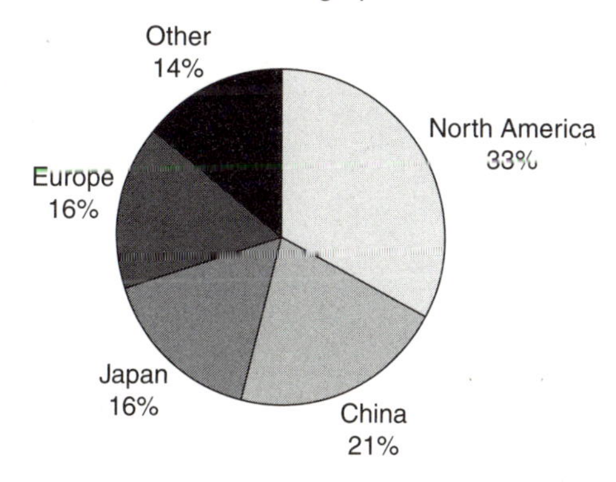

FIGURE 1.16 Palladium Global Geographic Demand
Source: Johnson Matthey public filings, 2010.

we seek as an extreme weather–based investor. The combination of a healthy demand story and a healthy supply-side story puts palladium into the Jackpot category.

We will cover the supply side and the specific investment opportunities in great detail in the chapter covering mine-based global climate shocks. The palladium-based investment opportunities we will cover include opportunities within the stock market, the futures market, and the ETF market.

Rare Earth Elements

The rare earth elements are typically defined as all of the elements in the periodic table that fall under the heading of the "lanthanoids" plus scandium and yttrium. These elements are shown in Table 1.2.

The rare earth elements are very unique in their supply/demand story. On the demand side, as shown in Figure 1.17, we see a globally growing and diverse set of applications.

On the supply side, which we will cover in more detail in the chapter covering mine-based global climate shocks, these materials are quite restricted currently. China is the dominant player and has instituted a severe curtailment on exports of rare earths. Their monopoly power is allowing

TABLE 1.2 Rare Earth Elements

Rare Earth Elements	Chemical Symbol
Scandium	Sc
Yttrium	Y
Lanthanum	La
Cerium	Ce
Praseodymium	Pr
Neodymium	Nd
Promethium	Pm
Samarium	Sm
Europium	Eu
Gadolinium	Gd
Terbium	Tb
Dysprosium	Dy
Holmium	Ho
Erbium	Er
Thulium	Tm
Ytterbium	Yb
Lutetium	Lu

Source: Periodic Table of the Elements.

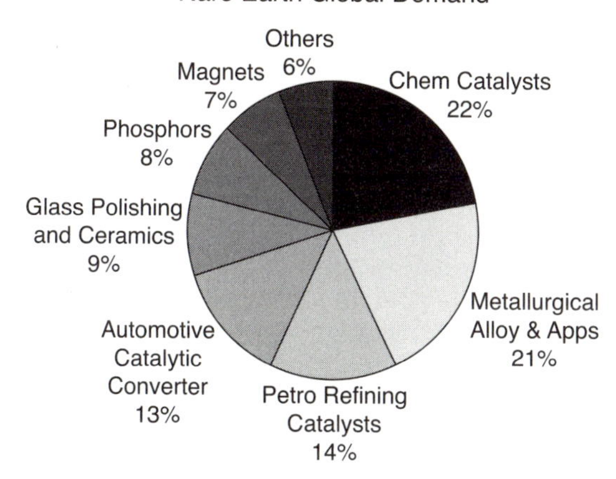

FIGURE 1.17 Rare Earth Global Demand
Source: USGS, 2010.

them to create a politically induced supply shock, thus keeping the rare earth market quite tight. The combination of a growing global demand but a tight supply story puts rare earths directly into the Jackpot category at this time.

As we will see in the chapter covering mine-based global climate shocks, there are investment opportunities in rare earths in the stock market and in the ETF market. Extreme weather–based investment opportunities within the corporate bond market and the foreign exchange market are generally nonexistent with regard to rare earths.

Potash Rock

Potash the rock and potash the downstream fertilizer share the same exciting end-market demand story (see the story for corn earlier in the chapter for additional detail). The end-market demand for potash rock is shown in Figure 1.18.

On the supply side, the potash source rock has a solid story. Not only is there a limited number of high-quality, low-cost sources of potash rock globally, but any new capacity additions are large and expensive, take a long time, and are quite visible. With limited capacity additions on the way, and with strong demand, potash falls into the Jackpot category. We will cover the supply side in more detail later.

As an extreme weather–based investor, there are numerous ways to invest in potash. We not only cover potash fertilizer investing in the chapter on "Global Climate Shock Number 3: Farmland Droughts, Floods, and

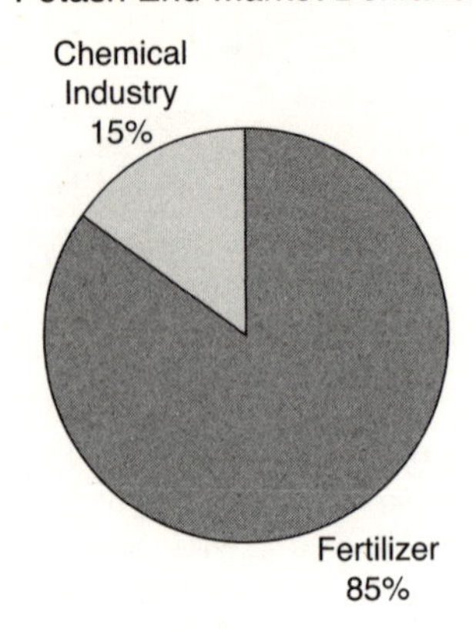

FIGURE 1.18 Potash End-Market Demand
Source: USGS, 2010.

Frost," but we also cover it in the chapter on "Global Climate Shock Number Two: Flooding Mines." In general, the investment opportunities are within the stock market and the corporate bond market. The nutrient potash does not trade in the futures market, nor are there any opportunities in the currency markets related to potash. On a related note, however, there are opportunities within the futures market for corn, as will be covered in detail.

NEUTRAL COMMODITIES

Zinc

Zinc ranks fourth in terms of world global metal production behind iron, aluminum, and copper. The end-market demand drivers for zinc are shown in Figure 1.19.

Globally, the demand for zinc is growing generally in line with GDP. This is an attractive feature of zinc.

The healthy demand curve for zinc, however, is partially negated by the supply-side story. Contrasting zinc with copper, for example, for which there are relatively few locations in the world with high-quality reserves, zinc is relatively more abundant, including 29 percent of global production coming out of the Chinese region, as shown in Table 1.3.

The heavy geographic concentration of zinc coming from the Chinese region is a negative attribute of this commodity. Do not confuse this with

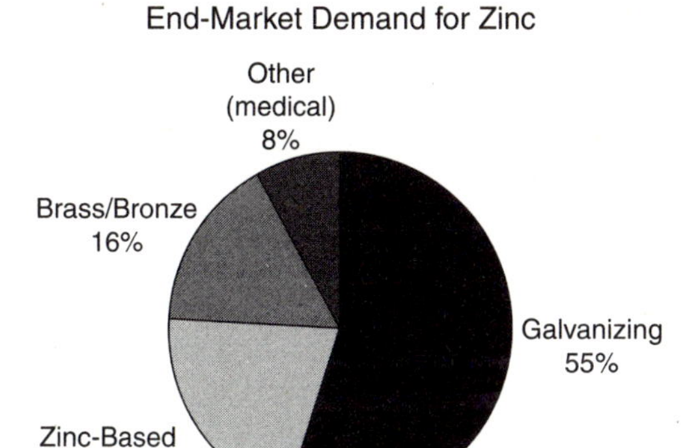

FIGURE 1.19 End-Market Demand for Zinc
Source: USGS, 2010.

the demand side. We love the fact that China has excellent and growing demand for zinc. However, we do not generally like it when a massive quantity of global supply comes from the Chinese region primarily because the incentive for increased employment for the very large population compels the government to continue to add capacity, even though it may not be the lowest cost capacity in the world. This tends to lead to overabundance of supply and relatively weak commodity pricing.

In terms of its attractiveness as a commodity for extreme weather–based investing, it generally falls short on multiple levels. As we

TABLE 1.3 Global Zinc Production

Global Production of Zinc	Percent Split
China	29%
Peru	13%
Ausi	12%
India	6%
United States	6%
Canada	6%
Mexico	5%
Kazakhstan	4%
Bolivia	4%
Ireland	3%
Other	12%
Total	**100%**

Source: USGS, 2010.

just talked about, despite the fact that demand is growing for the product globally, it is lacking in attractiveness on the supply side given its Chinese-based concentration and the potential for oversupply. This negative supply-side picture puts zinc into the "neutral" commodity box. This by itself does not necessarily negate extreme weather–based investment opportunities; however, the fact that there are very few "pure play" zinc producers (meaning all they produce is zinc) causes a zinc supply shock to be largely diluted and negated. We therefore do not spend much time on zinc as a primary target of extreme weather–based investing. There is, however, one company worth mentioning that would benefit from a supply shock–induced price spike in zinc, and that is India-based Vedanta (stock ticker VED LN). Zinc holds 21 percent of their total revenue and 55 percent of their operating income, according to recent Vedanta filings.

As you will see, aluminum metal production and steel production also share the same supply-side weakness as zinc, thus driving all three of them into the Neutral category.

Aluminum

Aluminum is derived from bauxite ore. It takes four tons of bauxite ore to make two tons of aluminum oxide via the "Bayer process" in a refinery. It further takes two tons of aluminum oxide to make one ton of aluminum metal in a smelter.

Aluminum metal has very desirable characteristics primarily related to its strength-to-weight ratio relative to other heavier materials such as steel. Its light weight characteristics allow it to take market share from other materials in the transportation sector where light weighting is important in order to conserve energy. Lighter cars, for example, get better gas mileage than heavier cars. This fact, in combination with rapid demand growth in the emerging regions of the world, make the demand side of aluminum look very attractive. The other key demand drivers for aluminum are shown in Figure 1.20.

The supply side of aluminum metal production partially negates the strong demand story. Approximately 40 percent of aluminum smelter production takes place in China. Similarly, China consumes approximately 40 percent of total global demand for aluminum. Given the very strong local demand for aluminum, China naturally wants to build aluminum production capacity to help satisfy local demand but also to increase the employment rate among its vast quantity of people. Very often, when China is at the helm in terms of being the leader in global capacity additions, the result is very often global oversupply. This negative supply-side dynamic pushes the aluminum metal commodity into the Neutral category. This is also the case with steel and zinc.

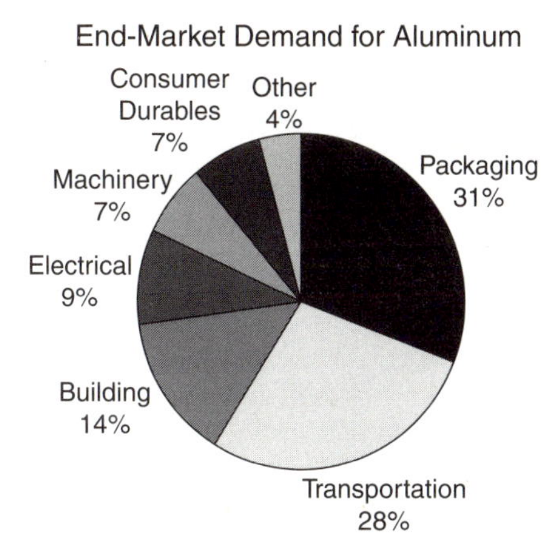

FIGURE 1.20 End-Market Demand for Aluminum
Source: USGS, 2010.

Despite the weaker supply-side story relative to its peers such as copper, there are still some extreme weather–based investing opportunities in the aluminum commodity, particularly within the futures market, but with the potential for some additional opportunities within the stock market and bond markets. Extreme weather–based investing opportunities within the foreign exchange market are nonexistent relative to this commodity. The specific investment opportunities will be covered in detail later.

Steel

Steel is a remarkable material in the sense that is represents the vast majority of global metal usage on a tonnage basis and yet it is far cheaper than other metals. Steel is used in structural applications, as shown in Figure 1.21, where its very high strength properties are required. The demand side of the supply/demand picture for steel is quite robust. Despite the weakness in demand that is occurring in the United States for steel, particularly within the nonresidential construction markets, the demand picture globally is quite strong as a result of the growth in the emerging regions of the world. Global steel demand runs at the rate of about 4 to 6 percent per year, according to figures cited at the U.S. Geologic Survey (USGS) web site, with China consuming roughly half of the total. So the demand side of steel looks very good, particularly when you consider that the United States currently is still in trough demand conditions and really just represents additional upside potential in global demand.

On the supply side of steel, China is also the dominant player, as shown in Table 1.4.

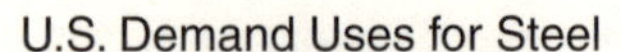

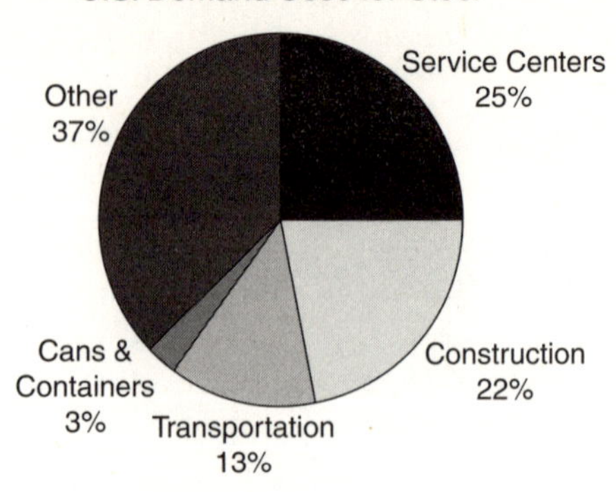

FIGURE 1.21 U.S. Demand Uses for Steel
Source: USGS, 2010.

Interestingly, steel, aluminum, and zinc all share a similar supply-side story, with China being the major global producer. Of course, China has a voracious appetite for all of these commodities, and therefore China is adding capacity in these products to help keep up with demand, but they also need to help their vast quantity of people stay employed. They have historically added capacity despite being a high-cost producer. This supply-side risk places steel into the Neutral category along with aluminum and zinc despite very healthy global demand.

TABLE 1.4 Global Raw Steel Production

Global Raw Steel Production	Percent Split
China	45%
Japan	8%
United States	6%
India	5%
Russia	5%
Korea	4%
Germany	3%
Brazil	2%
Ukraine	2%
France	1%
United Kingdom	1%
Other	18%
Total	**100%**

Source: USGS, 2010.

In terms of extreme weather–based investing opportunities in steel, unfortunately, there are very few for a couple of key reasons. Steel is not a mined product; it is a manufactured product derived from the combination of iron ore, coke (derived from metallurgical coal), and limestone. Therefore, the climate-based supply shock opportunities actually rest much more with the input raw materials rather than the steel itself because iron ore and metallurgical coal are both mined products.

What appears to be an excellent geographic concentration with Chinese-based steel production is actually quite fragmented, with dozens of steel makers scattered across the country, thus reducing the potential for a meaningful climate-based supply shock.

What about the steel futures contract? Even this path is not our most compelling weather-based investment opportunity. Earlier, we said that the key inputs to steel (i.e., iron ore and metallurgical coal) are more likely than steel itself to see a global climate shock because the key inputs are mined materials. So if we think that we should buy a steel futures contract in the event of a global climate shock in one of the two key steel-making raw materials, we are assuming that the cost increase of the raw material will eventually get passed through to steel prices, and therefore the futures contract investment would be successful. However, there tends to be a chronic oversupply in the global steel market due in part to the reasons mentioned earlier. When a market is oversupplied, the market tends to lose its pricing power and therefore may have some trouble increasing the price of steel sufficiently. If steel prices have trouble rising, then the futures contract investment would also struggle, thus making our opportunity in steel quite limited. Let's just say that more compelling opportunities exist elsewhere.

Nickel

A key end use for nickel is in the production of stainless steel because of the anticorrosion properties it imparts to the steel. This feature gets translated down into its key end-market demand drivers, which are shown in Figure 1.22.

On a global basis, the demand for nickel is quite healthy and is diversified enough such that it tends to grow with GDP globally.

On the supply side, nickel has experienced new capacity additions and is currently running at near decade-long highs in global supply of inventory. However, nickel meets our criteria for concentrated global geologic production and therefore will have opportunity in extreme weather–based investing. Given the weak supply and healthy demand backdrop, nickel falls into the Neutral category.

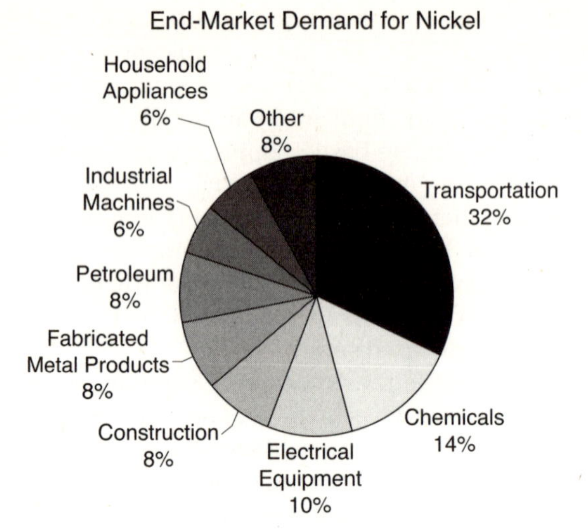

FIGURE 1.22 End-Market Demand for Nickel
Source: USGS, 2010.

Nickel has extreme weather–based investing opportunities in the futures market and a few select opportunities in the global stock market. We will discuss these opportunities in more detail in the chapter covering flooded mines.

Lead

Lead is an amazing material. It is functionally very versatile and effective and has the potential to be used in many diverse applications. However, its negative health effects offset its goodness. For this reason it has been outlawed in many applications, particularly in the United States, in such areas as leaded gasoline, paint, and piping, among others. Within the United States, the dominant application for lead is in the lead acid battery, which is the type of battery used in cars. The U.S.-based demand drivers are shown in Figure 1.23.

The fact that lead acid batteries are a key demand driver is good news globally because the global auto build rate, particularly in China, is very strong. However, the combination of the declining demand in many other global applications and the longer-term potential threat for lithium-based batteries taking share from lead acid batteries pushes the global demand picture for lead into the Neutral category.

On the supply side, lead is mined globally, with China producing by far the largest portion of the global total. Table 1.5 shows the global split-out in terms of global lead mining production.

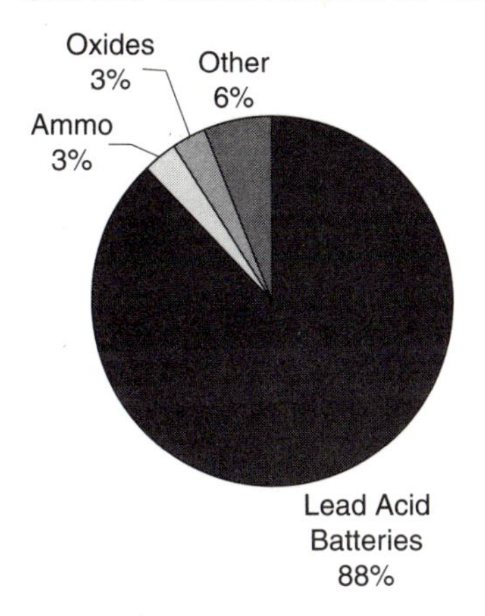

FIGURE 1.23 U.S. End-Market Demand for Lead
Source: USGS, 2010.

The supply side of lead suffers from the same situation we saw for steel, aluminum, and zinc in terms of its exposure to Chinese-based capacity additions. In addition to the Chinese supply-side exposure, lead also has an additional negative on the supply side. This metal has a remarkably high rate of successful recycling. Within the United States, the secondary

TABLE 1.5 Global Lead Production

Global Lead Production	Percent Split
China	43%
Australia	15%
United States	10%
Peru	7%
Mexico	5%
India	2%
Russia	2%
Bolivia	2%
Canada	2%
Sweden	2%
South Africa	1%
Ireland	1%
Poland	1%
Other	7%
Total	**100%**

Source: USGS, 2010.

supply (i.e., from scrap lead remelted) provided for 82 percent of total U.S. consumption of lead, according to the USGS. This is fairly easy to envision given that virtually every car battery gets recycled within the United States.

The combination of the mediocre global supply/demand situation puts lead into the category of Neutral.

When it comes to extreme weather–based investing, lead does not have the characteristics we like. Specifically, we will not keep lead a primary focus of extreme weather–based investing for the following reasons:

- Lead has a heavy concentration of its supply-side mining exposure to China, which has repeatedly been shown to contribute to overcapacity.
- The fact that a very meaningful portion of the supply side of lead comes from a secondary source (i.e., scrap recycled lead) does not lend itself to global climate shocks because the scrap material is not mined but, rather, derived from other sources such as from old car batteries.
- The number of public companies with lead as a major portion of their total revenue is quite limited, and therefore extreme weather–based opportunities in the stock market in this metal are quite limited.
- Lead does trade in the futures market, but other commodities present better opportunities, so for our purposes we will avoid this metal.

Thermal Coal

The global demand for thermal coal is growing. The end-market driver for thermal coal is almost exclusively for the production of electricity. With GDP in both China and India growing at very healthy rates, this implies the need for continued growth in the power-generation space, and, in fact, coal-based power generators are being constructed in China and India currently. The recent earthquake in Japan and its very negative impact on the nuclear power generators' reputation, contributes to increased growth in coal-based power plants. There are also substitution threats from other power sources, including natural gas, hydroelectric, wind, and solar, but given the very abundant coal reserves globally, including those found in China and the United States, coal will continue to play a very important role for the foreseeable future, including the expectation for positive global growth.

On the supply side, relative to metallurgical coal, thermal coal is one step less impressive. In the case of metallurgical coal, the regions of the world where one can find very high quality metallurgical coal are very limited and, in fact, are limited to three major regions, including the eastern United States, western Canada, and eastern Australia. Thermal coal, by contrast, has a wider distribution of fairly high quality material. It therefore gets put into the Neutral category, whereas metallurgical coal stays in the Jackpot category.

Because of the wider distribution of supply-side availability in thermal coal relative to metallurgical coal, we will keep our focus on metallurgical coal when it comes to identifying extreme weather–based investing opportunities.

As we will discuss in great detail, the extreme weather–based investing opportunities in metallurgical coal reside within the stock market and the corporate bond market. Opportunities within the futures market and the foreign exchange market are essentially nonexistent.

Soda Ash

Soda ash is a key ingredient in the production of glass. The most prevalent type of glass is what is known as soda-lime-silica glass. The *soda* refers to soda ash. Soda ash is also used for the production of detergent and other chemical uses, but glass is the dominant end market, as shown in Figure 1.24.

Globally, the demand for soda ash is a slightly positive number, in the 1 to 2 percent range despite the persistent weakness in the construction market in the United States. The construction market uses glass, which is the key end-market driver for soda ash.

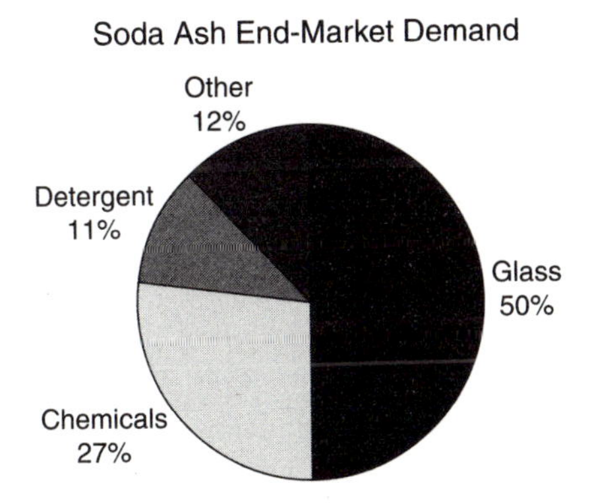

FIGURE 1.24 Soda Ash End-Market Demand
Source: USGS, 2010.

The supply side of soda ash is even more interesting. Generally speaking, there are two dominant ways to produce soda ash. It can be produced from natural trona rock or via a synthetic method. Table 1.6 shows the production split globally for soda ash.

The vast majority of the global, naturally mined production of soda ash comes from Green River, Wyoming. In this single mining complex, four major producers operate, the most important of which, from a public investing standpoint, is the FMC Corporation (stock ticker FMC). The "natural" method of production is the low-cost method globally. However, the low cost is referring to the cost of production at the mine. Unfortunately for the Wyoming producers, they export a large percentage of their soda ash, which roughly doubles their costs because they are so distant from the Asian region that receives much of their product, thus putting the synthetic soda ash makers in China right in line with the Wyoming producers on a delivered-cost basis.

Synthetic soda ash, by contrast, is made from salt and limestone, both of which are in abundant supply. The process is energy intensive as well, thus contributing toward its higher cost position relative to the natural soda ash at the mine.

China is the largest producer and consumer of soda ash. As a result of their desire to satisfy local demand, the relative availability of raw materials, and the desire to keep their vast population employed, there tends to be overcapacity in the region.

Given the mediocre global demand growth, in combination with the tendency for overbuilding of capacity, soda ash falls into the Neutral category.

From the standpoint of an extreme weather–based investor, the number of opportunities is fairly limited in soda ash. On the surface, it appears very promising because of the massive concentration of natural soda ash in Green River, Wyoming. In the event of an extreme weather occurrence at this mining complex, it indeed would represent a major supply shock to

TABLE 1.6 Global Soda Ash Production

Global Production of Soda Ash	Percent Split
Natural, United States (mostly Green River, WY)	22%
Synthetic	75%
Other	3%
Total	**100%**

Source: USGS, 2010.

the global soda ash balance; however, it would directly, negatively impact the major public producer of soda ash in that region, FMC Corporation. So, in the case of FMC Corporation, the guidance is simply to avoid this stock in the event of an extreme weather occurrence in Green River, Wyoming. Even when considering the eight major producers of soda ash in China, we cannot make money on them because the industry is state run. To further confound the problem, soda ash does not trade in the futures market and also offers no opportunities within the foreign exchange markets. The bottom-line guidance in soda ash for the extreme weather–based investor is avoiding FMC Corporation when an extreme weather event hits Green River, Wyoming.

BIG PROBLEM COMMODITIES

As shown earlier, most commodities fall into the category of Jackpot or at least Neutral in terms of their supply/demand fundamental outlook. One good example of a commodity that currently deviates from this trend is natural gas in North America. Although the demand side for natural gas has been and will continue to be generally mature but positive, the supply side has experienced a dramatic increase in the availability of natural gas in North America. Historically, the industry, when drilling for natural gas deposits in North America, has always drilled vertical holes. However, this creative industry decided to try drilling horizontally while hydro-fracturing the surrounding stone. This seemingly simple switch allowed the industry to generate a massive quantity of new supply of natural gas for years to come. Figure 1.25 shows the location of the major new supply, known for its geologic descriptor, "shale plays."

Figure 1.26 shows the location of the more traditional sources of supply within what is known as the "tight gas plays."

This massive new supply is obviously bad news for the natural gas producers because it has put downward pressure on the price of natural gas.

From the standpoint of an extreme weather–based investor, despite the fact that natural gas falls into the Big Problem commodity list, there are still some opportunities, particularly in relation to hurricanes in the United States. We will talk about this in detail in the chapter covering extreme hurricane events. Specifically, there are investment opportunities within the stock market, the bond market, and the futures market. We also cover real-life examples and results in the natural gas space in Chapter 8, "Real-Life Examples: Execution, Results, and Timing."

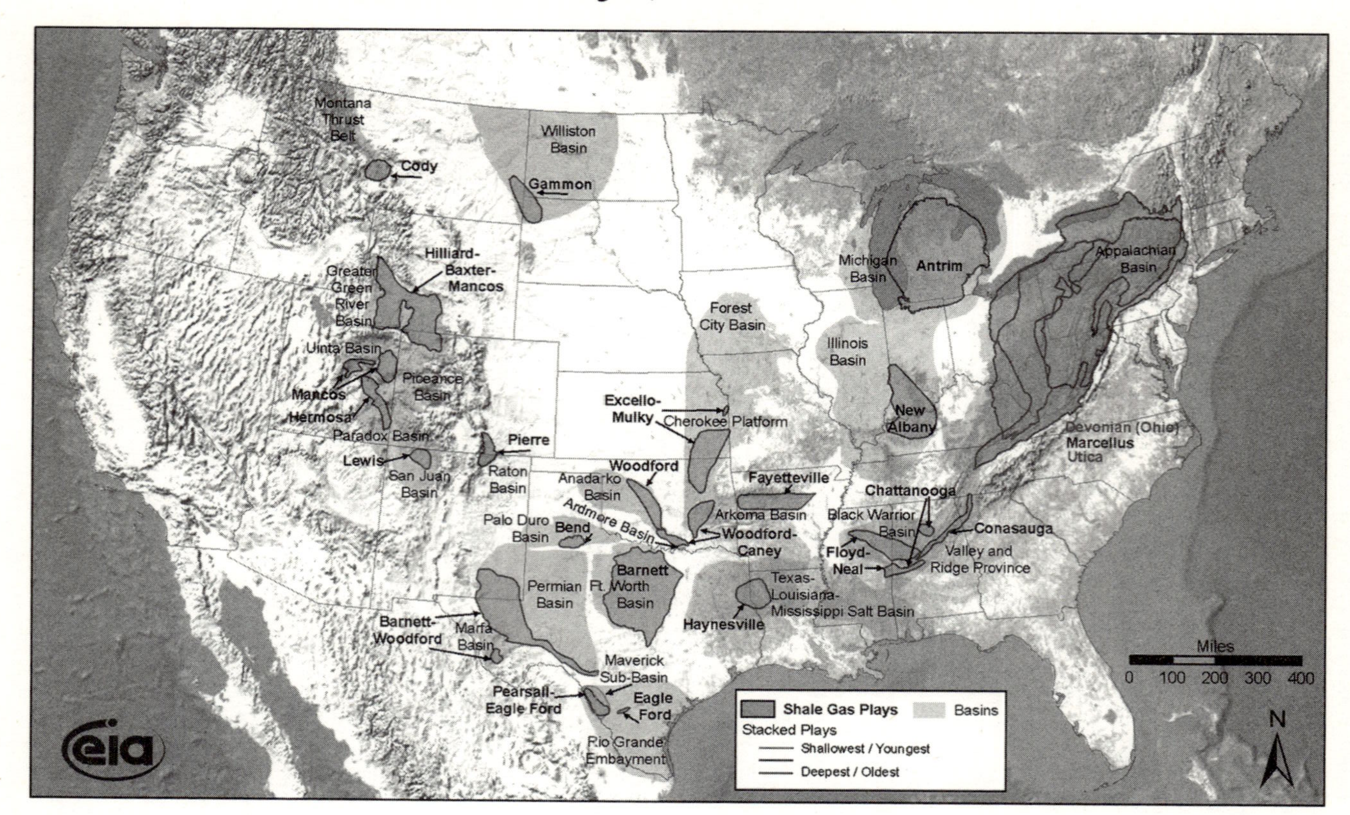

FIGURE 1.25 Shale Gas Plays
Source: EIA, 2009.

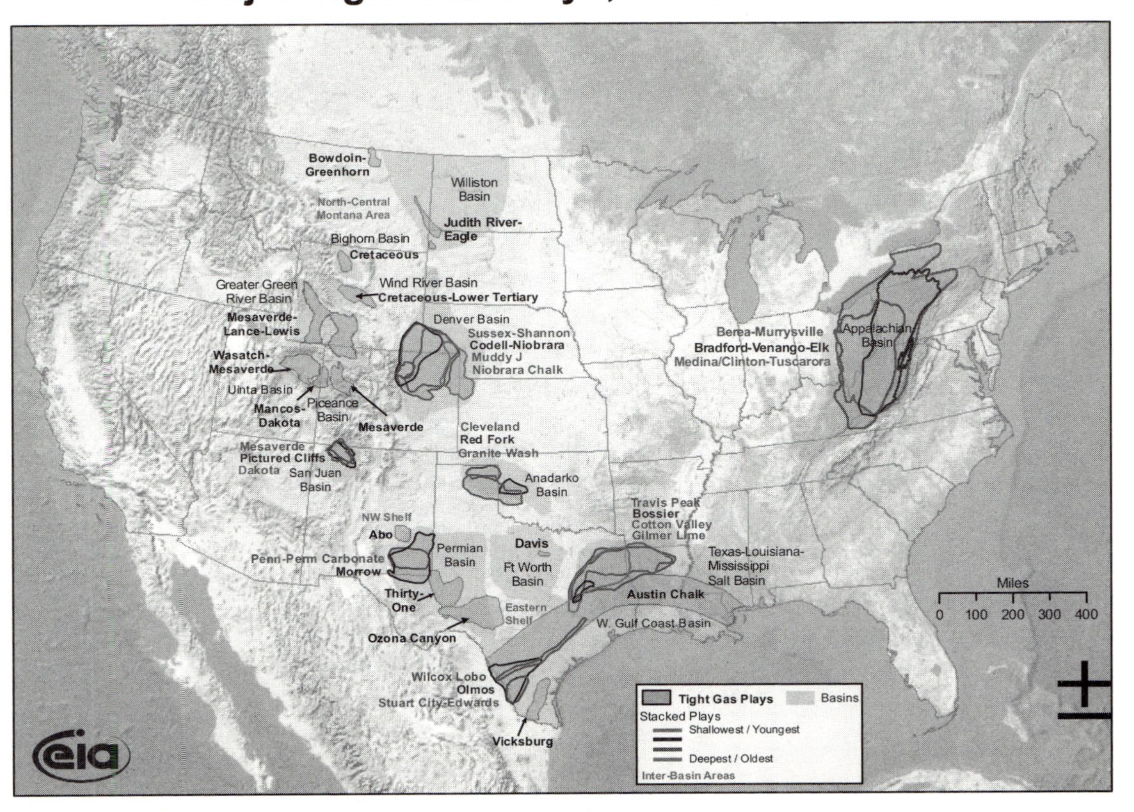

FIGURE 1.26 Majcr Tight Gas Plays
Source: EIA, 2009.

OTHER LIMITED OPPORTUNITY COMMODITIES

When considering investment opportunities associated with global climate shocks, a key aspect of the analysis is the likelihood of a meaningful supply-side disruption. In the next section we consider lean hogs and live cattle.

Lean Hogs and Live Cattle

When considering whether lean hogs or live cattle could have a meaningful supply-side disruption based on the weather, it appears unlikely and therefore negates meaningful opportunities for extreme weather–based investing. Granted, it is true that supply shock can and will occur in meat-producing operations, but this is more often related to disease and not the weather.

Another indirect investing opportunity within the meat-producing industries is related to a supply shock in the price of corn. A supply shock in the price of corn increases the cost of feeding the hogs and cattle, but this is a second-tier effect, which is far better exploited by investing in corn futures directly.

CHAPTER 2

Where to Invest: Stocks, Bonds, or Futures?

As we just saw in the last chapter, the vast majority of commodities have a very positive to at minimum neutral fundamental outlook. This fact alone makes the commodity sector very attractive from an investor's standpoint. However, when you throw into the mix the ever-increasing supply shocks that will occur as a result of global climate change, the commodity sector becomes downright exciting. These two items act synergistically. Supply shocks by themselves cause the price of commodities to move upward, but when the supply/demand story of a commodity is already tight and market participants are already filled with fear of not having adequate supply, the emotions behind the pricing dramatically move upward from fear to panic, thus causing market prices to climb more sharply, allowing investors increased opportunities to make money.

Later, we will begin to dig into the specific global climate changes and the corresponding supply shocks that will occur. We will then identify the names of the specific securities that will benefit from the supply shock and methods on how to make money in the name, but first, let's take an introductory look at each of the financial markets and see how they compare against each other as we begin to explore how we will invest in global climate shocks and extreme weather events.

Table 2.1, in a very simplistic format, outlines the investing opportunities associated with global climate shocks and extreme weather events. It is put in a grid format, with each of the major financial markets across the top row and each of the relevant commodities and sectors listed along the left column.

TABLE 2.1 Extreme Weather Investing Opportunities

Extreme Weather Investing Opportunities?	Stocks	Bonds	Futures Market	Foreign Exchange
Metal/Mining/Oil/Gas	Many	Many	Many	A few
Agriculture/Grains	Many	Many	Many	Very little
Petrochemical	A few	A few	Very little	Very little
General Industrial	A few	A few	Very little	Very little

As shown in the body of the grid, each combination is assigned a qualitative descriptor summarizing the sheer quantity of investing opportunities with regard to global climate shocks and extreme weather events. As shown in the table, there are many investing opportunities in the grid, thus making investing in global climate shocks a very exciting and diversified way of making money. As shown, however, some sectors offer more opportunities than others. *It is really up to the reader to decide in which financial market and in which commodity or sector to make his or her investments. In this chapter of the book, we emphasize the stock market simply as the easiest for all of the reasons mentioned later, but by no means does this negate the vast opportunities outlined in the other financial markets. After reading through all of the numerous investing opportunities available, the reader will naturally migrate toward certain sectors depending on his or her comfort and skill level.*

For many readers, the stock market will be the market of choice for the following basic reasons:

- It is the market where the general public has their best understanding.
- It is the market where anyone can buy and sell stocks in his or her own personal trading account.
- It is a highly liquid market where there are numerous buyers and sellers, thus making it very easy to make a trade.
- Finally, there is a high degree of correlation between the stocks, bonds, and futures markets, so why bother getting involved in the bond and futures market when you can get the same basic result in an easier-to-handle stock market.

The rest of this chapter will go on to prove the correlation among these three markets (stocks, bonds, and futures). If you have already decided that the stock market is your market of choice, then feel free to skip the remainder of this chapter.

As noted in the title of this book, our goal is translate extreme weather events into opportunities in the stock, bond, and futures markets. The tie that binds these three large markets together is the core fundamentals of the business. We will use the commodity copper to make this point.

The fundamentals of copper are very strong. On the demand side, the world needs copper. In fact, we need more and more copper every year. This is occurring despite the fact that the U.S. construction market has been weak over the past couple of years. This points to the strong demand for copper in emerging economies in places such as China and India, which more than compensate for the lack of demand in the mature economies in the recessionary period. The strong global demand will get even better once the construction market in the United States and other mature regions recovers. So, basically, the demand side of the copper market looks very good.

On the supply side, there are some projects on the table to add capacity in the major copper-producing regions of the world, including Chile and Peru, but these projects will barely keep up and possibly fall short of demand, particularly because the copper that we are mining is gradually yielding less and less copper metal per ton of ore.

So when comparing the demand projection versus the supply projection, the market is expected to remain fairly tight, meaning possibly not enough copper to go around. If you are a buyer of copper and there is not enough copper to go around, you are more than willing, because of your fear, to pay extra for your copper to guarantee that you get your supplies. These fears cause the price of copper to go higher. When copper prices go higher, the producers of copper become very happy. Higher copper prices translate into much stronger profits (i.e., earnings) for the company. Correspondingly, with higher profits come higher stock prices. If we use the company Freeport McMoRan (ticker FCX) as an example of a company that produces copper, this relationship becomes obvious. The stock market is keenly aware of the price of copper. As shown in Figure 2.1, the price of copper and the stock price of Freeport McMoRan are highly correlated. Simply put, when the price of copper goes up, the stock price of Freeport McMoRan goes up, too.

Now in the case of Freeport McMoRan, the correlation between the price of copper and the stock price of the company is very strong because copper represents about 75 percent of their total revenues (i.e., sales). The company also makes gold and molybdenum (affectionately known as "moly" in the industry), but these other two materials are a smaller part of the total, so copper is the key driver for this stock. In order to get the biggest bang for your buck in stock picking, it is absolutely critical to have at your fingertips the names of the companies that are most affected by each commodity. We will cover this critical point in detail later.

Now, what about the relationship between stocks and bonds? As mentioned earlier, the tie that binds the stock, bond, and futures markets is

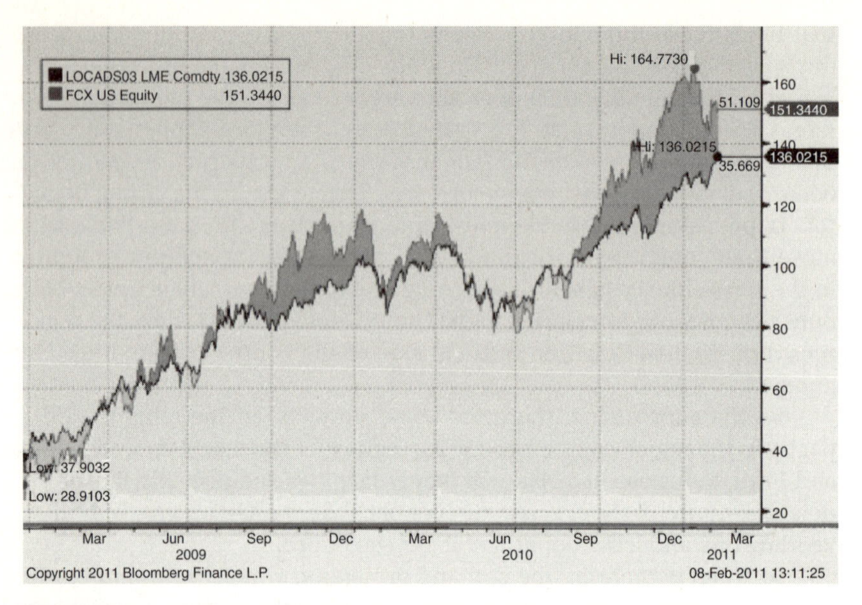

FIGURE 2.1 Copper Price versus Freeport Stock Price (ticker FCX)
Source: Used with permission of Bloomberg Newswire Permissions.

the fundamentals of the business. If the outlook and price of copper are favorable, then the stock market responds favorably, but what about the bond market? The bond market is very similar in the sense that bonds also like strong fundamentals of a business. Just a few notes on corporate bonds will help clarify. Corporate bonds trade from investor to investor just like stocks. Corporate bonds also have a price at which investors buy and sell just like stocks. When a company is improving and becoming less risky, the corporate bond price goes higher, and when a company is getting worse and therefore more risky, the bond price gets lower. A picture is worth a thousand words, so let's look at the graph in Figure 2.2. As you can see, the graph shows how stocks and bond prices tracked each other with a high degree of correlation over the past four years, before and after the great global recession. The Standard & Poor's (S&P) 500 market index was used to represent the stock market and the High Yield Corporate Bond index was used to represent the bond market. This is excellent news from our perspective because this means when deciding on how to invest in companies that are affected by global warming and global watering, you can choose whichever market makes the most sense for your particular needs.

One key difference between the stock and bond market is that if you own a corporate bond, you get the added benefit of a fixed cash payment

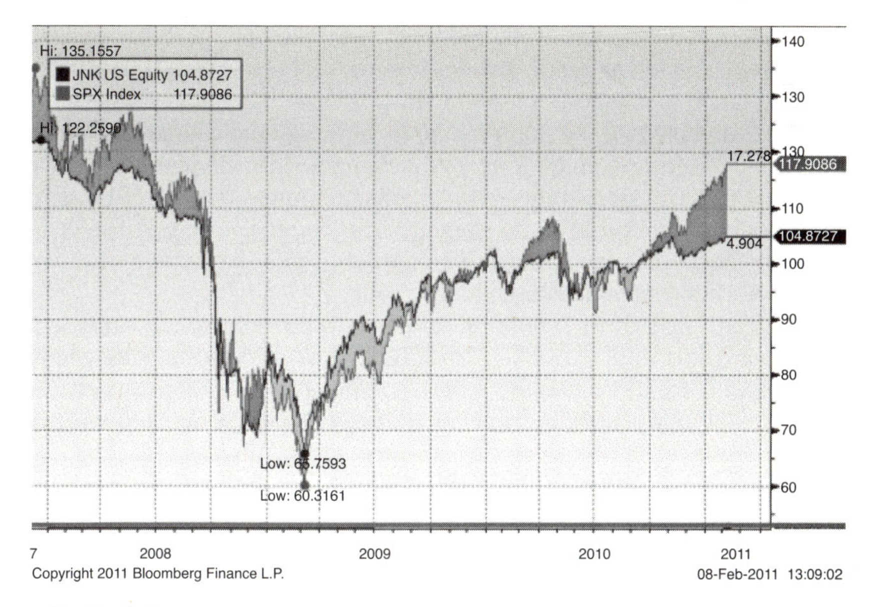

FIGURE 2.2 S&P 500 Stock Index and High Yield Corporate Bond Index
Source: Used with permission of Bloomberg Newswire Permissions.

from the company on a semiannual basis. This cash payment on the bonds is actually the interest the company pays on its debt. The company's debt is the same thing as the bond that you buy in the bond market. The beauty of it is that even if the price of the bond doesn't rise at all, you will still receive your cash interest payment every six months. Of course, you can always buy a stock of a company that pays a dividend and get the same basic benefit.

What about the futures market? (See Chapter 13, "Basic Principles of Futures Market Investing" for greater detail on the futures market.) We will use copper as our example. The beauty of the futures market rests in its simplicity. Its primary focus is on the price of copper. When the spot price (the price you pay for immediate delivery of a commodity) of copper goes higher, then the futures market price of copper shifts higher as well. You can buy copper in the futures market, and then if the market price of copper skyrockets, you win. More specifically, you are allowed to buy a copper futures contract and sell the same contract if copper prices skyrocket and then take the profit, without having to lay your fingers on a single pound of copper in the transaction. The futures market doesn't care if the chief financial officer of a company retires. It doesn't care whether a company's pension is underfunded. It doesn't care that a company lays off 1,000 workers. It just cares about the price of copper. The stock market and corporate bond market, by contrast, care more about these miscellaneous items, thus making the futures market very attractive. However, the futures market is most often used by the professional-type investors and is not generally used by the public at large, although the general public can also enter the futures market as an investment via their personal trading account broker or indirectly via the exchange-traded fund (ETF) market, which will be covered later. So, depending on the type of investor you are, you can choose to enter the futures market to take advantage of the extreme weather events discussed in this book.

Now let's drill down a little bit farther into the comparison among the stock, bond, and futures markets for those investors who are willing to expand beyond the more familiar stock market. As was shown earlier, over the three- to four-year time frame, including before, during, and after the global recession, it was clear that all three major markets tracked each other quite closely. Part of the reason why is that during the great recession, all markets got hit so hard they had to be correlated because everything was going down—and going down hard. Likewise, during the recovery phase beginning in March 2009, it was fairly easy to continue the strong correlation among markets because all markets were at bottom and had nowhere to go but up. However, what does the comparison look like if we strictly look at the full-year 2010 time frame after the markets had already partly recovered?

To get a solid picture of all three markets in parallel, 11 random commodities were selected, as shown in Table 2.2.

For each commodity, a representative publicly traded company was selected that is a major producer of that commodity. Both the stocks and the bonds of that particular company were used in the comparison. This method allows us to see the direct relationship among these three financial markets with very little room for error because the stock and bond market price responses were linked to identical business conditions. In addition, because we chose the full year 2010 as our reference time frame, we are less impacted by the drastic trough in the global recession that occurred in early 2009. By eliminating the 2008 and 2009 time frame from the analysis, we remove a contributor to the higher-than-average market correlation that existed over this time frame.

Now it is time for the results. We can slice and dice the table a number of ways, but basically the key conclusions among the three major markets are:

- As shown, if you were involved in any of the three markets in 2010, you would have done quite well.
- Even the generally more conservative bond market saw a median value return in 2010 of about 12 percent—not too shabby. Part of the reason

TABLE 2.2 Results for 2010: Stocks versus Bonds versus Futures

2010 RESULTS				
Commodity	**Representative Company**	**Stocks**	**Bonds**	**Futures**
Copper	Freeport McMoRan	**44%**	10%	29%
Aluminum	Alcoa	−5%	**12%**	9%
Gold	Newmont	26%	13%	**30%**
Silver	First Majestic Silver	**250%**	NA	76%
Platinum	Northam Platinum	−6%	NA	**21%**
Met Coal	Teck Resources	**61%**	12%	17%
Corn	Potash Corp	37%	12%	**38%**
Oil	Hess	**21%**	13%	12%
Steel	Steel Dynamics	0%	8%	**34%**
Thermal Coal	Peabody	34%	14%	**59%**
Natural Gas	Chesapeake	−8%	**10%**	−30%
	Winner Count	**4**	**2**	**5**
	Median Value Return	**26%**	**12%**	**29%**
	Mean Value Return	**41%**	**11%**	**27%**

Note: Spot price used for thermal and met coal.
Source: Public pricing data.

for such a strong bond market was that we were and currently are still in gradual recovery mode from the trough of the recession.

- As shown in the "winner count," the biggest winner based on median value return for 2010 was the futures market, which won 5 times out of 11 total commodities. Notice that if we used the average value instead of the median value, the stock market would have been the overall winner. I chose to use the median value because, as you can see, the 250 percent stock return for First Majestic Silver was somewhat of an anomaly in the results, but, importantly, does demonstrate the serious upside potential when investing in the stock market.
- One might wonder how the futures market might be able to beat the stock market in general. The explanation is quite simple. When there is a global climate shock or extreme weather event, the price of the affected commodity will go up, and therefore the price of that commodity in the futures market will also go up. The stock price of a producing company will also very likely go up, but isn't it possible that the company could have some other negative issue occurring at the same time? The company may be seeing its oil-based energy costs skyrocketing. So even though the price of the commodity that they make is going up, their profit margins may not rise quite as quickly due to cost inflation. This is the primary reason why the futures market often will

win out over the stock market. This is also why the futures market can be so lucrative for an extreme weather–based investor.

- The corporate bond market results were quite impressive as returns go. Bond market returns are made up of two parts. Part one is the coupon payment (i.e., interest payment) that the corporate bond investor receives semiannually, and part two is the appreciation (or depreciation) of the dollar price of the bond. The 12 percent median return is good but, generally speaking, typically will be lower than stock market returns. This is expected because a basic tenet of investing is that the higher the risk, the higher should be the reward. You may ask why the risk of a stock is higher than the risk of a bond. The basic reason rests in the capital structure pecking order. In theory, in the event of a bankruptcy, the senior secured debt holders get paid in full first. They are followed in order by the unsecured bondholders in the pecking order. Last in the soup line are the equity holders, who, in the event of bankruptcy of the firm, may end up with none of their original investment.
- Also, bond prices in general have a property called asymmetry. This means that the amount of upside potential in the price of a bond is quite limited compared to its downside. However, the bonds are higher up in the waterfall pecking order and therefore are generally a less risky investment as compared to the stock market. In addition, if a company's earnings are outstanding and growing rapidly, the debt market can't reap all of that benefit because, after all, the bond is actually a debt instrument and at the end of its life will very likely get refinanced at its original price of 100 cents on the dollar. So most of the upside value must go predominantly to the stockholder, which also translates into more volatile pricing for stocks than for bonds.
- The key question we must decide on is do we want to miss out on some of this upside potential by avoiding the stock and futures markets altogether and instead stick with the more stable (but with decent return) bond market? The answer to that question is a personal decision depending on the needs, comfort, and skill level of the investor. After reading through all of the chapters in this book, you will be in a much better position to decide which financial market is best for you when it comes to extreme weather–based investing.
- A key point to keep in mind when you see the "action plan" tables throughout this book is that even though the table refers to a company by name and by stock ticker, the table is also saying that the corporate bond market for that particular company is also a very viable option.

As you will see, the next five chapters are a very exciting tour around the globe. In these chapters we will identify the key extreme weather events. In addition and critically important, after reading these chapters, you will then have right at your fingertips the key companies and commodities to target for investment in virtually any extreme weather situation around the world.

CHAPTER 3

Global Climate Shock Number One: Excess Snow and Ice

I have really good news for you. The next time you are shoveling back-breaking snow because your section of the world has been covered by the blizzard of the century, you can smile because you will know which stock wins under this scenario. Read on.

When it comes to excessive snow and ice, the effect can be a supply shock because of the difficulties of transporting various materials through these icy conditions, but in this case the more obvious direct impact is to increase the demand for certain items. Specifically, when extreme snow and ice conditions hit various regions, the demand for salt and snow blowers tends to rise.

SALT

In order to make money on a shift upward in the demand for salt we want to find not only the names of companies that are heavy in salt production to allow the biggest bang for the salt buck, but we also want to find companies that have publicly traded stock. Looking globally, the number one salt producer in the world is K+S Ag. They recently expanded their number one global salt position by acquiring the Morton Salt business from Dow Chemical after Dow acquired the salt business from Rohm and Haas. The other major public salt maker is Compass Minerals. K+S is more of a global player that has dominant positions in Germany and other Euro regions, as well as other regions throughout the world. Compass, by contrast, is more

TABLE 3.1 Salt Makers' Revenue Mix

Product Revenue Split	K+S Ag (Ticker SDF Gr)	Compass Minerals (Ticker CMP)
Salt	30%	87%
Fertilizer	67%	13%
Other	3%	0%
Total	**100%**	**100%**

Source: Public filings.

of a North American player. Tables 3.1 and 3.2 show the split-out of their revenue and geographic positioning.

From the standpoint of taking advantage of a step change increase in the demand for salt, Compass Minerals is the winner because of its heavy concentration in this product. They are nearly a "pure play" in salt. Looking geographically, the comparison is shown in Table 3.2.

The geography is important to keep in mind because in the event of horrendous snow and ice storms in China, Compass Minerals would not see a direct benefit. However, because 75 percent of their revenue is located within the United States, they would benefit directly when heavy snow and ice hits the United States. In fact, the most recent winter, ending in early 2011, resulted in plenty of snow and ice hitting the United States. Correspondingly, the stock price for the key "salt play," Compass Minerals (ticker CMP), saw an impressive upward movement easily outperforming the Standard & Poor's (S&P) 500 Index. We cover this stock reaction as well as a number of other specific real-life examples in Chapter 8, "Real-Life Examples: Execution, Results, and Timing."

TABLE 3.2 Salt Makers' Geographic Positioning

Geographic Revenue Split	K+S Ag	Compass Minerals
Germany	18%	
Other European Regions	36%	
United States		75%
Canada		20%
Rest of World	46%	5%
Total	100%	100%

Source: Public filings.

SNOW BLOWERS

Ask anyone in the northeastern United States why they chose to shovel snow instead of using a snow blower and you might hear comments such as, "Well, I think it's good exercise" or maybe "Snow blowers are too expensive." However, after shoveling snow for the fifth time in one winter, with each storm averaging more than a foot of snow, the average diehard snow shoveler is guaranteed to be considering the transition to buying their first snow blower. After all, shoveling snow, in fact, is not good for your health given the very awkward and asymmetric movements and strain on your back, not to mention heart problems for the people who go from a daily sedentary position directly into the grueling snow shoveling position in one fell swoop.

With this backdrop, the other investing opportunity in the excessive snow and ice category is snow blowers or snow removal vehicles. Within this category there are three major public companies. They include Toro (ticker TTC), Oshkosh (ticker OSK), and Honda (ticker HMC). Unfortunately, this category suffers from product diversity and brand awareness away from snow removal. Honda and Oshkosh, in particular, are known for vehicles other than their snow removal products. Toro, by contrast, although also very diversified in product line, is fairly well recognized for its snow removal product line as a meaningful piece of its brand. Therefore, Toro would be the recommended snow/ice play within the snow blower category, primarily for its headline relationship to winter weather. Comparatively, however, the Compass Mineral salt play discussed earlier is the much preferred snow/ice play as a result of the extreme revenue concentration on salt in their product mix, particularly within the United States.

CHAPTER 4

Global Climate Shock Number Two: Flooding Mines

Severe regional mine flooding can cause a supply disruption to many different commodity materials. Let's first focus on the mining operations that meet two criteria: (1) where the supply/demand fundamentals are quite favorable over the long term and (2) where there is a high concentration of high-quality ore in a limited number of geographies. By satisfying these two criteria, we can hone in on the companies that will see the strongest reaction to a supply shock associated with mine flooding. Before we drill down on specific ore types and specific ore producers, it is critical to understand some basic concepts related to supply shocks and commodity reactions. These key points are outlined here:

- A supply shock that is induced by nature (i.e., global warming and global watering) in the commodity world almost exclusively causes what is called a negative supply change. In other words, the availability of supply gets lower. It doesn't really matter what the cause, whether it's from rain, snow, ice, drought, hurricane, earthquake, volcano, or mine flooding, the result is the same: a temporary—or even permanent, in some cases—reduction in the available supply of the commodity. As we talked about earlier, when the supply becomes limited, in the commodity world, the price of that commodity moves higher and very often quite sharply. So, for the most part, the supply shocks associated with global warming and global watering are great for commodities in general (see next bullet for exception).
- The exception to the benefit to commodity prices from a supply shock is to the company or region that saw the supply shock. For example,

the region that gets wiped out by flooding obviously does not benefit from the supply shock because they have no product to sell. However, all other regions and producers that were not impacted by the supply shock benefit greatly.

- It is therefore critically important as an investor to have at the tip of your fingers the geographic breakdown of all critical commodities and all of the associated producers of those commodities.
- With this information at hand, you can at a moment's notice take action on the appropriate companies that were not involved in the major supply shock but at the same time will benefit from the supply shock.

So, again, the two critical criteria that we ideally want to meet are (1) having excellent supply/demand fundamentals (i.e., remember the list of Jackpot commodities that we covered earlier) and (2) having concentrations of the ore in a limited number of geographies.

It is important to realize that extreme weather–based investing is ideal with a commodity where the supply/demand balance is tight (i.e., Jackpot commodity) because the market is already in fear of a shortage of the commodity. However, that does not mean that extreme weather–based investing opportunities do not exist in commodities with only mediocre or even poor supply/demand fundamentals. We can make money even with these less attractive commodities because the supply shock from the weather event causes the tightness in the supply/demand balance and therefore leads to an increase in the price of the commodities. We will cover both the Jackpot and non-Jackpot commodities in the following sections.

METALLURGICAL ("MET") COAL

We will start with metallurgical ("met") coal because it provides a very relevant example: at the time of this writing, eastern Australia is under water from flooding, and this material meets both of the key criteria mentioned earlier. As a reminder, it has very strong core supply/demand fundamentals, as discussed earlier, and there is only a limited number of geographies around the globe that produce very high quality metallurgical coal (aka hard coking coal).

Looking geographically first, the best met coal in the world is produced in eastern Australia, the eastern United States, and western Canada. Combined, these three countries accounted for approximately 80 percent of the globally traded coking coal supply in 2010. Of that 80 percent, eastern Australia accounted for 64 percent of the total; the United States, 24 percent; and Canada, 12 percent.

TABLE 4.1 Higher-Quality Met Coal Producers

Company	Met Coal % of Sales	Australia	Eastern United States	Western Canada	Other
BHP	12%	X			
Teck	50%			X	
Xstrata	15%	X			X
Anglo	12%	X			X
Walter	80%		X	X	
Alpha	13%		X		
Rio Tinto	10%	X			X
MacArthur	94%	X			
Peabody	10%	X			

Source: Public filings.

In terms of demand for this material, China consumes roughly half of world production of coking coal. China also produces a great deal of their own coking coal as well, but the quality and quantity are insufficient, hence the need for imports to satisfy domestic demand. Simply by looking at the geographic split on the data, it becomes immediately obvious how serious it is that eastern Australia is under water from flooding. This represents a massive supply shock to the global met coal market. Knowing that Australia is in trouble from a supply standpoint, we want to identify the met coal producers that are not in Australia because they are the ones that will benefit as the price of met coal increases. In addition, the companies that will benefit the most from this supply shock are the ones that have a larger portion of their total revenue coming from met coal. The largest high-quality met coal producers are shown in Table 4.1. The public equity ticker symbol for each of these companies is shown in Table 4.2.

TABLE 4.2 Equity Tickers for Met Coal Producers

Met Coal Company	Stock Price Ticker
BHP	BHP Au
Teck Resources	TCK
Xstrata	XTA Ln
Anglo	AAL Ln
Walter	WLT
Alpha	ANR
Rio Tinto	RIO Ln
MacArthur	MCC Au
Peabody	BTU

Source: Public filings.

As can be seen, BHP, Xstrata, Anglo American, Rio Tinto, and MacArthur will not be the biggest beneficiaries in this case because their dominant met coal operations are located in eastern Australia, and therefore their production of met coal is impaired with the floods. The companies that will benefit the most will be non-Australia-based and also the companies with very high concentrations of met coal as a percentage of their total revenue. The companies that meet this criterion are Teck Resources, Walter Energy, and Alpha Natural Resources. Simply by looking at the geographic data in combination with the company data in the table, we can immediately decide which companies benefit the most from flooding in any region globally.

For explanatory purposes, let's look at another hypothetical situation. If, for example, flooding took place in the western Canada met coal mines, who would be the biggest losers and who would be the biggest winners? Try it, based on the information in the tables. The biggest losers would be Teck Resources and Walter Energy because their mines are located in western Canada. The biggest winner would be MacArthur because it's solely in eastern Australia and it's a coal pure play (i.e., the only product it makes is coal). Did you get those answers? Good, I assume you did because having access to the table makes the answer very clear. The second-tier winners in this example would be all of the other non-Canadian players, including BHP, Xstrata, Anglo, Rio, and Alpha. These would be considered

TABLE 4.3 Action Plan Table for "Met" Coal

	Where's the Flood? Eastern Australia	**Where's the Flood? Western Canada**	**Where's the Flood? Eastern United States**
Biggest Winners (Ranked)	Walter Energy (80%) Tech Resources (50%) Alpha Natural (13%)	MacArthur (94%) BHP (12%) Xstrata (15%) Alpha (13%) Anglo (12%) Rio Tinto (10%) Peabody (10%)	MacArthur (94%) Tech Resources (50%) Xstrata (15%) BHP (12%) Anglo American (12%) Rio Tinto (10%) Peabody (10%)
Biggest Losers (Ranked)	MacArthur (94%) Xstrata (15%) BHP (12%) Anglo American (12%) Rio Tinto (10%)	Walter Energy (80%) Tech Resources (50%)	Walter Energy (80%) Alpha Natural (13%)

Note: Arcelor and Peabody have offered to acquire MacArthur at the time of this writing.
Source: Public filings.

only a second-tier win because met coal represents a smaller amount of their total revenue and therefore would have a smaller but still meaningful positive impact on its stock price.

What appears obvious is only obvious because we now have access to the geographic and product split tables. With this information at our fingertips, we are now prepared to take immediate action when we get word of a global mine flood in any geography in the world.

Table 4.3 summarizes our action plan when met coal mines get flooded in each key global geography.

For a detailed review of actual stock reactions, as well as execution strategies for this particular flooding event in eastern Australia, see Chapter 8, "Real-Life Examples: Execution, Results, and Timing."

IRON ORE

When looking at the potential mine flood impact on the iron ore market, we will use the same methodology we used for met coal. We will first identify the major geographies where iron ore is produced and then identify the key players in the market and decide who would be the greatest winners and losers in various circumstances. Starting with the global geographic split for the production of iron ore, we arrive at Table 4.4.

As shown in Table 4.4, the key geographies for iron ore production are Western Australia, China, Brazil, India, and the Commonwealth of Independent States (CIS). Because nearly half of the world's *steel* is produced in China, this means about half of total demand for *iron ore* comes from China as well. Similar to the met coal discussion, the quality and quantity

TABLE 4.4 Global Iron Ore Production

Iron Ore Production	Percent
Western Australia	24%
China	22%
Brazil	18%
India	13%
Commonwealth of Independent States	11%
Africa	4%
Canada	3%
Other	5%
Total Global Production	**100%**

Source: USGS, 2010.

of iron ore made in China is insufficient, and therefore they import iron ore from the geographies shown in Table 4.4. Mine flooding (or any other natural disaster for that matter) in any of these key regions would be highly disruptive to the iron ore world market. In particular, if western Australia or Brazil were impacted, the supply shock would be severe. In the event that China itself was the source of the troubles, the benefit to the iron ore market would be muted because the steel production could possibly be impaired as well, which would hurt the demand side for iron ore. For this reason, the opportunity for a beneficial supply shock to the iron ore market rests more so in western Australia and Brazil.

Now let's take a look at the key players in the iron ore market and decide who would be the biggest winners and losers under various circumstances. The iron ore market is one of the most impressive markets in the world in terms of consolidated oligopolies. Market participants tend to focus on what is called the "seaborne iron ore market." The reason this is the primary area of focus is that so much of the total market gets exported from Brazil and western Australia to China. Within the seaborne market, three major players dominate the world. BHP, Rio Tinto, and Vale together hold about 80 percent of the market. This fact, in combination with growing global demand from steel and relatively tight supply over the next couple of years, makes this market very attractive, as discussed earlier.

When we look at the top six public producers of iron ore and map out their geography and their weighting in iron ore relative to their total revenue, we obtain Table 4.5. The public equity ticker symbol for each of these companies is shown in Table 4.6.

Table 4.5 actually represents a detailed road map on how to react to news of global mine flooding. We can immediately not only determine who will be the winners and losers but also be able to rank them. For example, in the event that western Australia finds itself under water (which is

TABLE 4.5 Iron Ore Producers

	BHP	Rio Tinto	Vale	FMG	Vedanta	Cliffs
Iron Ore % of Total Revenue	**21%**	**28%**	**61%**	**100%**	**15%**	**89%**
Iron Ore geographic split:						
Western Australia	X	X		X		
Brazil			X			
India					X	
Other						X (United States)

Source: Public filings.

TABLE 4.6 Equity Tickers for Iron Ore Producers

Iron Ore Company	Stock Price Ticker
BHP	BHP Au
Rio Tinto	RIO Ln
Vale	VALE3 Bz
FMG (Fortescue)	FMG Au
Vedanta	VED Ln
Cleveland Cliffs	CLF

Source: Public filings.

exactly what is occurring in eastern Australia at the time of the writing of this book), which geographies and companies will win, and which will lose? Obviously, the companies in western Australia will lose because, as discussed earlier, the companies directly hit by a supply shock do not win. This means companies like BHP, Rio Tinto, and FMG would lose in this case. In terms of rankings, FMG would lose the most because they have 100 percent of their total company revenue tied up in iron ore and more specifically 100 percent tied up in western Australia. The winners in this scenario would be Cliffs Natural Resources, Vale, and Vedanta, in that order. Cliffs would be the biggest winner because they have 89 percent of their revenue in iron ore and none in western Australia. Similarly, Vale would also do quite well, with 61 percent of their total revenue in iron ore. Finally, Vedanta would partially benefit because the price of iron ore would rise, but Vedanta would benefit less than its peers because iron ore represents only 15 percent of its total revenue.

Table 4.7 summarizes our action plan when iron ore mines get flooded in each key global geography.

TABLE 4.7 Iron Ore Action Plan Table

	Where's the Flood? Western Australia	Where's the Flood? Brazil	Where's the Flood? India
Biggest Winners (Ranked)	Cliffs Natural (89%) Vale (61%) Vedanta (15%)	FMG (100%) Cliffs Natural (89%) Rio Tinto (28%) BHP (21%) Vedanta (15%)	FMG (100%) Cliffs Natural (89%) Vale (61%) Rio Tinto (28%) BHP (21%)
Biggest Losers (Ranked)	FMG (100%) Rio Tinto (28%) BHP (21%)	Vale (61%)	Vedanta (15%)

Source: Public filings.

COPPER

Our next area of focus will be copper mining. Copper mining meets the two key criteria we set out earlier: (1) it has a solid fundamental supply/demand outlook, and (2) it has geographic concentration in at least one region such that in the event of a supply shock in any concentrated region, clear winners and losers would emerge.

Let's start by looking at the geography of copper mining, as shown in Table 4.8.

As shown, Chile is the clear geographic leader in terms of copper mining. Likewise, Codelco holds the number one global market share in copper mining, but this is a private company owned by the Chilean government and so will not be included in our discussion. (However, it should be noted that Codelco does, in fact, have corporate bonds outstanding that provide a low yielding but relatively low risk return). The rest of the copper production is rather diversified, so the opportunities for major supply shocks are restricted to the Chilean region predominantly. Even within Chile the mines are distributed along the entire length of Chile, which is as long as the United States is wide. However, Chile is all coastal in nature, which represents an extreme weather–based threat. Even if a portion of Chile suffered from excessive flooding (or earthquakes, for that matter), it would represent a significant supply shock.

We will next map the major copper producers against their geography and their relative exposure to copper, as shown in Table 4.9. The public equity ticker symbol for each of these companies is shown in Table 4.10.

TABLE 4.8 Global Copper Mining Production

Copper Mine Production	Percent
Chile	34%
Peru	8%
United States	8%
China	7%
Indonesia	6%
Australia	5%
Russia	4%
Zambia	4%
Canada	3%
Other	21%
Total Global Production	**100%**

Source: USGS, 2010.

TABLE 4.9 Copper Producers

	Freeport	BHP	Xstrata	Anglo	RioTinto	Southern Copper
Copper % of Total Revenue	80%	13%	40%	39%	14%	70%
Copper Mine Geographic Split:						
Chile	X	X	X	X	X	
Peru		X	X	X	X	X
United States	X				X	
Indonesia	X				X	
Australia			X		X	
Other			X		X	X

Source: Public filings.

As shown, the companies that are most levered to copper are Freeport McMoRan and Southern Copper, both having copper hold more than 70 percent of their total revenue. Xstrata and Anglo American are the second-level companies in terms of leverage to copper, with nearly half of their revenue coming from this metal. Copper is interesting from a supply shock analysis perspective because almost all players have some mining activity occurring in Chile, the geographic leader. So, in the event that Chile was out of commission for copper production, all producers would suffer, with the biggest winner being Southern Copper, given its Peruvian-based copper. Conversely, if the Peruvian copper mines were out of commission, the biggest loser clearly would be Southern Copper. The fact that nearly all of the players have their hands in the largest geography makes copper more attractive from the futures market perspective. This is so because in the event that the entire Chilean coast was under water (or suffering from an earthquake, given that this region is traditionally very dry from a climate perspective), it would hurt the vast majority of the copper-based

TABLE 4.10 Equity Tickers for Copper Makers

Copper Mine Company	Stock Price Ticker
Freeport	FCX
BHP	BHP Au
Xstrata	XTA Ln
Anglo	AAL Ln
RioTinto	RIO Ln
Southern Copper	SCCO

Source: Public filings.

TABLE 4.11 Copper Action Plan Table

	Where's the Flood? Chile	Where's the Flood? Peru
Biggest Winners (Ranked)	Copper Futures Market Southern Copper (70%) Freeport McMoRan (80%)	Copper Futures Market Freeport McMoRan (80%)
Biggest Losers (Ranked)	Anglo American (39%) Xstrata (40%) Rio Tinto (14%) BHP (13%)	Southern Copper (70%) Anglo American (39%) Xstrata (40%) BHP (13%) Rio Tinto (14%)

Source: Public filings.

stock market and bond market, but the futures price of copper would skyrocket. The more geographically diversified copper makers would partially benefit from this scenario as well, however.

Table 4.11 summarizes our action plan when copper mines get flooded in each key global geography.

SILVER

Similar to the other mining sector discussions, we want to identify not only the regions of the world with the highest concentration of silver production but also the key silver producers that have publicly traded equity. Table 4.12 shows the regional concentration of silver production. As shown,

TABLE 4.12 Global Silver Production

Global Silver Production	Percent Split
Mexico	17%
Peru	16%
China	13%
Australia	8%
Chile	6%
Bolivia	6%
United States	5%
Poland	5%
Russia	5%
Other	19%
World	**100%**

Source: USGS, 2010.

TABLE 4.13 Silver Producers

Silver Producer	Silver % of Sales	Mexico	Peru	Ausi	Bolivia	United States	Poland
BHP	3%			X			
Fresnillo PLC	53%	X					
KGHM	13%						X
Pan American	50%		X				
Coeur d'Alene	69%	X		X	X	X	
First Majestic	100%	X					
Southern Copper	6%	X	X				
Hecla Mining	51%	X				X	

Source: Public filings.

Mexico, Peru, and China are the dominant geographies, representing nearly half of total global production.

Therefore, these are the key geographies to watch when it comes to extreme weather events such as mine flooding. Recall also from earlier discussions that, technically speaking, any supply-side shock to a concentrated geography will cause commodity prices to go up. This is true whether it's due to mine flooding, earthquakes, or politically based supply disruptions. For the purpose of this discussion, we will speak in terms of mine flooding in these key regions. Now that we know the location of the key geographies, we can map them against the publicly traded silver-producing companies and their respective production geographies, as shown in Table 4.13. The stock ticker symbols for these companies are shown in Table 4.14.

Now that we know which companies have exposure to the highly concentrated Mexico and Peru-based silver-producing regions, we are in a

TABLE 4.14 Equity Tickers for Silver Producers

Silver Company	Stock Price Ticker
BHP	BHP Au
Fresnillo PLC	FNLPK
First Majestic	AG
KGHM	KGHPF
Pan American	PAA Cn
Hecla Mining	HL
Southern Copper	SCCO
Coeur d'Alene Mine	CDM Cn

Source: Public filings.

TABLE 4.15 Silver Action Plan Table

	Where's the Flood? Mexico	Where's the Flood? Peru	Where's the Flood? China
Biggest Winners (Ranked)	Pan Am. (50%) KGHM (13%) BHP (3%)	First Maj. (100%) Coeur d'Alene (69%) Fresnillo (53%) Hecla (51%) KGHM (13%) BHP (3%)	First Maj. (100%) Coeur (69%) Fresnillo (53%) Hecla (51%) Pan American (50%) KGHM (13%) Southern Cop. (6%)
Biggest Losers (Ranked)	First Maj. (100%) Coeur (69%) Fresnillo (53%) Hecla Mining (51%) Southern Cop. (6%)	Pan American (50%) Southern Cop. (6%)	Local Chinese Silver Miner

Source: Public filings.

position to act. Our action plan for responding to extreme weather–based climate events impacting the silver market can be summarized in Table 4.15.

ALUMINUM

Recall from the "commodity stories" segment that the demand for aluminum metal is fast growing globally but that the China-based oversupply risk puts this commodity into the Neutral category. Nevertheless, there are still some extreme weather–based investing opportunities.

It all starts with the bauxite ore, the key material used to produce aluminum metal. I have good news and bad news. The good news is that bauxite ore meets our extreme weather–based investing criteria because it possesses the desirable heavy geographic concentration in a limited number of locations around the world. Table 4.16 shows key global geographies of bauxite production.

The bad news is that bauxite is not a commodity that trades in the futures market. So in the event of a global climate shock in the key bauxite-producing regions of Australia, China, or Brazil, it will not provide a direct means of investing in the futures market. However, there are indirect methods that we can use. When bauxite is in short supply due to a global climate event, prices will rise. As a result of the spike in the price of bauxite, the price of downstream aluminum metal will also rise as the cost-push dynamic gets pushed through the system. It is this indirect method in which

TABLE 4.16 Global Bauxite Production

Global Bauxite Production	Percent Split
Australia	34%
China	19%
Brazil	15%
India	9%
Guinea	8%
Jamaica	4%
Kazakhstan	3%
Russia	2%
Other	2%
Suriname	1%
Venezuela	1%
Greece	1%
Guyana	1%
Total	**100%**

Source: USGS, 2010.

we can enter the futures market via an aluminum metal contract (see Chapter 13 on how to invest in the futures market). The fact that bauxite is only an indirect investment opportunity may be a blessing in disguise, however. Because of the lag effect on downstream aluminum metal pricing, it may actually buy us some time. In other words, if bauxite ore sees an extreme weather event that drives up the price of bauxite, it may be the best time to enter the aluminum metal futures price contract because the price of the metal will not spike as quickly as the price spike in the bauxite ore. Interestingly, it may act as a leading indicator for the aluminum futures market.

What about opportunities in the stock market? Extreme weather–based opportunities within the stock market for aluminum metal are even more challenging. The benefit, by contrast, in entering the futures market is that when metal prices go up, we win—period. However, in the case of the stock market, the aluminum producers, particularly the ones that do not make their own bauxite, see their key raw material input costs go up when bauxite prices rise. So the big benefit to the aluminum makers when the aluminum metal price goes up gets wiped away because their bauxite-based input costs spike upward as well. For this reason, our opportunities within the stock market when bauxite prices rise are less compelling.

What about the corporate bond market? As discussed earlier, very often, publicly traded commodity-based companies provide us opportunities to invest not only in their public stock but also in their corporate bonds. As shown earlier, the stock and bond markets are very similar in the sense that they both respond positively to improvements in company profits and they both respond negatively to a squeeze in profits when raw material prices (like bauxite) spike. Because of this correlation with the stock market, our opportunities within the bond market when bauxite prices rise are also less compelling.

NICKEL

Despite falling into the Neutral category, there are still some extreme weather–based investment opportunities with the commodity nickel. Nickel meets the general criteria for extreme weather–based investing given its desirable geographic concentration of production in a few select areas of the globe. The global split in terms of nickel production is shown in Table 4.17.

As shown in Table 4.17, the top nickel-producing regions of the globe are Russia, Indonesia, the Philippines, and Canada. In the event the nickel-producing regions of these countries see an extreme weather event, the price of nickel will rise. The most direct method of playing this investment is in the nickel futures market because the futures market does not care which nickel-producing geography sees the extreme climate event. It benefits from any extreme climate event globally as long as it's within one of the key nickel-producing regions. Therein lies the beauty of the futures

TABLE 4.17 Global Nickel Production

Global Nickel Production	Percent Split
Russia	17%
Indonesia	15%
Philippines	10%
Canada	10%
Australia	9%
New Caledonia	9%
China	5%
Cuba	5%
Colombia	5%
Brazil	4%
South Africa	3%
Botswana	2%
Other	6%
Total	**100%**

Source: USGS, 2010.

TABLE 4.18 Nickel Producers

Nickel Producers	Nickel % of Total Revenue
Norilsk	43%
Eramet	27%
Xstrata	9%
Vale	8%
BHP	4%

Source: Public filings.

market. Recall, however, that because nickel does not fall into the Jackpot category but rather the Neutral category, its upside potential in the futures market is one step lower in attractiveness as compared to Jackpot commodities such as met coal, iron ore, or copper.

Within the stock market, the geography of the extreme climate event is critical because it defines who are the winners and who are the losers. When looking around the globe at the publicly traded producers of nickel with a meaningful quantity of nickel in their total portfolio, one must search far and wide. The five primary publicly traded nickel producers meeting this criterion are shown in Table 4.18. The percentage next to each name represents the percent of total revenue held by its nickel division. The stock ticker symbols of these companies are shown in Table 4.19.

The ticker for Norilsk shows the letters RU at the end, which refers to Russia. The Eramet ticker ends with FP, which refers to France. The BHP ticker ends with AU, and this refers to Australia. The ticker for Vale ends with BZ, which refers to Brazil. The ticker for Xstrata ends with LN, which refers to London. So, in other words, the available stock plays are from the far corners of the earth. To be more specific, the nickel mining assets for each player are defined in Table 4.20. These locations help us define who will be the winners and who will be the losers in the event of a global climate shock within each region.

TABLE 4.19 Equity Tickers for Nickel Producers

Company	Global Ticker
Norilsk	GMKN RU
Eramet	ERA FP
BHP	BHP AU
Xstrata	XTA LN
Vale	Vale3 BZ

Source: Public filings.

TABLE 4.20 Nickel Producers

	Norilsk	Eramet	BHP	Xstrata	Vale
Nickel % of Total Revenue	43%	27%	4%	9%	8%
Mining Assets Location:					
Russia	X				
South Africa	X				
Australia			X	X	
Columbia			X		
New Caledonia		X		X	
Norway				X	
Canada				X	X
Dominican Republic				X	
Brazil				X	
Indonesia					X

Source: Public filings.

Now that we have the key players and key nickel-producing geographies at our fingertips, it becomes very easy to map out an action plan in the event of a global climate shock in the nickel market. Our action plan for the global nickel market is shown in Table 4.21.

The biggest global winners and losers are mapped out in the table. Importantly, notice that regardless of the location of the global climate event, the nickel futures market is always the winner because, as we pointed out earlier, the futures market does not care about which geography has the global climate shock, it simply likes when any geography in the world is impacted. It is also important to note that although the focus of the table—and the book, for that matter—is on extreme weather events, the table also applies to other similar supply shocks, which may result from strikes,

TABLE 4.21 Nickel Action Plan

	Where's the Flood? Russia	Where's the Flood? Indonesia	Where's the Flood? Canada
Biggest Winners (Ranked)	Nickel Futures Eramet (27%) Xstrata (9%) Vale (8%) BHP (4%)	Nickel Futures Norilsk (43%) Eramet (27%) Xstrata (9%) BHP (4%)	Nickel Futures Norilsk (43%) Eramet (27%) BHP (4%)
Biggest Losers (Ranked)	Norilsk (43%)	Vale (8%)	Vale (8%) Xstrata (9%)

Source: Public filings.

earthquakes, politically based supply disruptions, and a host of other possible events whose end result is a supply shock.

PLATINUM

Platinum may well be my favorite weather-based investment story. It has all of the key characteristics we look for as an extreme weather–based investor, including the following:

- The demand story is strong globally. The commodity fills a true need and has a growth story not only within the emerging regions, but even within the developed regions of the world.
- The supply side is highly geographically concentrated. This is very, very important. As we will see, global production of platinum is located in just a few very large handfuls geographically around the globe.
- Not only is most of the global production in highly concentrated pockets geographically, but these specific geographies also have the potential for politically rooted supply shocks. It does not really matter what causes the supply shock. The focus of this book is extreme weather–based investing opportunities, but if a supply shock is brought on by a political issue, the price of the impacted commodity will go up regardless. Commodities have no biases. (See Chapter 8 for a concise list of the rules associated with extreme weather–based investing.)

Now let's take a closer look at the supply situation in platinum. Table 4.22 outlines the geographic global production split for platinum.

TABLE 4.22 Global Platinum Production

Global Platinum Production	Percent Split
South Africa	76%
Russia	13%
Zimbabwe	5%
Canada	3%
United States	2%
Other	1%
Total	**100%**

Source: USGS, 2010.

As shown, South Africa represents a whopping 76 percent of total platinum mine production. This is a staggeringly high number. Most commodities pale in comparison to this level of geographic concentration. To put this into perspective, as discussed elsewhere, corn and cocoa are also impressive in terms of their geographic production concentration. Corn had 41 percent of global production made predominantly in the Corn Belt of the United States. Cocoa had nearly 40 percent of global production coming out of the Cote d'Ivoire region, which is located in Northwest Africa. These two commodities are quite impressive from the standpoint of an extreme weather–based investor due to their high level of geographic concentration; however, they both pale in comparison to the platinum story. The 76 percent of global platinum production in the South African region is roughly double that found for corn and cocoa, hence the excitement around platinum.

Now let's take a look at the public companies that produce platinum. Table 4.23 shows each public company as well as the percentage of total revenue for that company coming from platinum sales.

As shown in Table 4.23, there are many publicly traded companies in the platinum space. In addition, as shown, the percentage of total revenue coming from platinum is impressively high. I say impressively because as an extreme weather–based investor, we like high concentration of revenue—as will be shown, it helps to amplify its position as the biggest loser or biggest winner depending on the geography of the global climate shock. As a point of reference, Table 4.24 depicts the stock ticker symbol for each of these companies.

Notice that the stock price tickers, as we have seen multiple times before, are often foreign based. Remember, we are dealing with *global*

TABLE 4.23 Platinum Producers

Company	% of Total Revenue From Platinum
Anglo American Platinum	64%
Johnson Matthey	74% (from precious metals)
Impala Platinum	77%
Lonmin PLC	68%
Stillwater Mining	47%
Northam Platinum	100%
Zimplats Holdings	58%
Royal Bafokeng Platinum	100%
Aquarius Platinum	85%
Norilsk	10%

Source: Public filings.

TABLE 4.24 Equity Tickers for Platinum Producers

Company	Stock Price Ticker
Anglo American Platinum	AMS SJ
Impala Platinum	IMP SJ
Johnson Matthey	JMAT LN
Lonmin PLC	LMI SJ
Stillwater Mining	SWC
Northam Platinum	NHM SJ
Zimplats Holdings	ZIM AU
Royal Bafokeng Platinum	RBP SJ
Aquarius Platinum	AQP AU
Norilsk	GMKN RU

Source: Public filings.

climate shocks, and for this reason we had better be ready to take advantage of global investing opportunities even if that means investing in companies from a foreign region, although ADR local equivalent stock tickers are available in many cases.

Now let's take a deeper dive into each company and find out where their mining assets reside globally. Table 4.25 is critical for our understanding of who will become the biggest winners and who will become the biggest losers depending on the location of the global climate event.

Now that we have this table at hand it becomes very easy to determine who will become the winners and losers. Examples always help to simplify things. What happens if the South African region sees a global climate event

TABLE 4.25 Platinum Producers

Company	Platinum Sales	South Africa	Russia	Montana (United States)
Anglo American Platinum	64%	X		
Impala Platinum	77%	X		
Lonmin PLC	68%	X		
Stillwater Mining	47%			X
Northam Platinum	100%	X		
Zimplats Holdings	58%	X		
Royal Bafokeng Platinum	100%	X		
Norilsk	10%	X	X	
Aquarius Platinum	85%	X		
Johnson Matthey*	74%			

*Downstream processor.
Source: Public filings.

(particularly at the Bushveld Complex in South Africa, the largest mining complex) that blocks off the flow of platinum out of the region? Try it. By looking at the table, who wins under this situation? Stillwater Mining would be the biggest winner. Good, I knew you would get that one. There is also an interesting addition to the table, as you can see. Johnson Matthey, although not one of the big miners in the South African region, does process and refine platinum and, in fact, has 74 percent of its total revenue coming from the precious metals group. In addition, its stock price over the past decade has had a remarkably high correlation with platinum prices. Therefore, as you will see below, we have included Johnson Matthey into the "Action Plan" table (Table 4.26) given its formidable position in the global platinum markets. Notice also that Johnson Matthey is not included in any of the "biggest loser" scenarios because it will generally benefit from higher platinum prices regardless of the location of a global climate shock.

OK, now let's do the inverse. What happens if the high-yielding mining complex in Montana sees a global climate event that blocks the flow of platinum out of that region—who loses and who wins? Stillwater Mining is obviously the biggest loser in this scenario. The biggest winners are the

TABLE 4.26 Platinum Action Plan

	Where's the Flood? South Africa	Where's the Flood? Montana (United States)	Where's the Flood? Russia
Biggest Winners (Ranked)	Platinum Futures ETFs J. Matthey (74%) Stillwater (47%)	Platinum Futures ETFs Northern (100%) Royal (100%) Aquarius (85%) Impala (77%) J. Matthey (74%) Lonmin (68%) Anglo (64%) Zimplats (58%)	Platinum Futures ETFs Northern (100%) Royal (100%) Aquarius (85%) Impala (77%) J. Matthey (74%) Lonmin (68%) Anglo (64%) Zimplats (58%)
Biggest Losers (Ranked)	Northern (100%) Royal (100%) Aquarius (85%) Impala (77%) Lonmin (68%) Anglo (64%) Zimplats (58%)	Stillwater (47%)	Norilsk (10%)

Source: Public filings.

remaining players around the globe, including all of the South African–based miners as well as Norilsk, centered in Russia. We can actually take this analysis one step further by ranking the winners and losers depending on the geography of the global climate event and also based on the percent of total revenue coming from platinum sales, as shown in Table 4.26.

At this point you might be wondering why "platinum futures" and exchange-traded funds (ETFs) in platinum are ranked higher than all of the publicly traded stock winners. The futures market and the ETF market do not care where the global climate event occurs. All they care about is the price of platinum. It makes life a bit easier if you do not have to worry about which names in the stock market are the winners and losers, but instead simply focus on the extreme weather events and the price of platinum.

Our final platinum-based action plan is shown in Table 4.26. With this table at your fingertips, you are now ready to take swift action in the platinum markets when an extreme weather event hits the defined region shown in the table.

PALLADIUM

As mentioned earlier, palladium is one of the platinum group metals (PGMs). Geologically speaking, this means that all of these metals, which include platinum, palladium, rhodium, ruthenium, and iridium, generally occur in nature together in the same mine site. The proportion of the mix coming out of the mine varies, however. This variation in the PGM product mix calls for a separate but similar discussion on palladium. However, given that platinum usually represents the bulk of the mix coming out of these mine sites and given that the same producers that make platinum are generally the same producers that make palladium, the reader can skip this section if desired and simply focus on the platinum opportunities. Nevertheless, we will dive into palladium because of its nuance of difference from platinum and because it has completely independent futures and ETF contracts available for investment.

Let's start with the supply-side global split within the palladium market. Table 4.27 shows the result.

As shown, the geographic split includes the same basic names as with the platinum global production split with the exception that Russia and South Africa switched positions and now Russia has the top spot. We would expect the same basic regions to be included in the table because of the coproduct nature of platinum and palladium.

TABLE 4.27 Global Palladium Production

Global Palladium Production	**Percent Split**
Russia	44%
South Africa	37%
Montana United States	6%
Canada	5%
Zimbabwe	3%
Other	5%
Total	**100%**

Source: USGS, 2010.

Now let's take a look at the global producers of palladium. Table 4.28 shows the key producers of palladium, along with the percentage of their total revenue that is derived from palladium sales. The public equity ticker symbol for each of these companies is shown in Table 4.29.

Notice from Table 4.28 that palladium represents a much smaller portion of total revenue for these companies than did platinum. This makes sense since platinum is typically found in higher proportion than palladium within the PGM series. There are some exceptions to this, including at the Stillwater Mining Company, where the ratio of palladium to platinum appears much higher than the rest of the peer group. When we dig down and look at the global geographic split for each player, the only difference we find is that we eliminate Northam Platinum and Royal Bafokeng from the table because of the negligible quantity of palladium reported in their mix, as shown in Table 4.30.

TABLE 4.28 Palladium Producers

Company	**% of Total Revenue from Palladium**
Anglo American Platinum	11%
Impala Platinum	13%
Lonmin PLC	9%
Stillwater Mining	53%
Northam Platinum	0%
Zimplats Holdings	13%
Aquarius Platinum	15%
Royal Bafokeng Platinum	0%
Norilsk	9%

Source: Public filings.

TABLE 4.29 Equity Tickers for Palladium Producers

Company	Stock Price Ticker
Anglo American Platinum	AMS SJ
Impala Platinum	IMP SJ
Lonmin PLC	LMI SJ
Stillwater Mining	SWC
Zimplats Holdings	ZIM AU
Aquarius Platinum	AQP AU
Norilsk	GMKN RU

Source: Public filings.

Using the same methodology we did with platinum, we can obtain a ranking of investment opportunities, depending on the location of the global climate event. Also included in the table are the opportunities associated with palladium futures and the ETF market. Notice that regardless of the location of the extreme weather event, the futures and ETF markets are ranked higher than the stock market opportunities. The beauty of the futures and ETF markets is that they focus only on the price of palladium. So, in the event of a global climate shock, the price of palladium goes up regardless of which key geography was hit, thus making them a very attractive alternative to the stock market.

Our final palladium-based action plan is shown in Table 4.31. With this table at your fingertips, you are now ready to take swift action in the palladium markets when an extreme weather event hits the defined region shown in the table.

TABLE 4.30 Palladium Producers

Company	Palladium Sales	South Africa	Russia	Montana (United States)
Anglo American Platinum	11%	X		
Impala Platinum	13%	X		
Lonmin PLC	9%	X		
Stillwater Mining	53%			X
Zimplats Holdings	13%	X		
Aquarius Platinum	15%	X		
Norilsk	9%	X	X	

Source: Public filings.

TABLE 4.31 Palladium Action Plan

	Where's the Flood? South Africa	Where's the Flood? Montana (United States)	Where's the Flood? Russia
Biggest Winners (Ranked)	Palladium Futures ETFs Stillwater (53%)	Palladium Futures ETFs Aquarius (15%) Impala (13%) Zimplats (13%) Anglo (11%) Lonmin (9%)	Palladium Futures ETFs Stillwater (53%) Aquarius (15%) Impala (13%) Zimplats (13%) Anglo (11%)
Biggest Losers (Ranked)	Aquarius (15%) Impala (13%) Anglo (11%) Lonmin (9%)	Stillwater (53%)	Norilsk (9%)

Source: Public filings.

RARE EARTH ELEMENTS

As we discussed in the section on commodity stories, rare earths at this time fall into the Jackpot category. In addition, they also meet the extreme weather–based investing criteria of geographic concentration. Let's drill down a bit further now into the supply side. Table 4.32 shows the global geographic split in mine production of rare earths.

As shown, China is the leader, by far, in terms of mined production of rare earths. In fact, Baotou City, and specifically the Baiyunebo Mine located in Inner Mongolia, is considered to be the world capital of rare earths. In the event of an extreme weather event at this location, rare earth prices would climb.

Given their monopoly status, China recently decided to significantly reduce the allowable exports to other regions around the globe. This book

TABLE 4.32 Global Rare Earth Production

Rare Earth Global Mine Production	Percent Split
China	97.3%
India	2.0%
Brazil	0.4%
Malaysia	0.3%
Total	**100%**

Source: USGS, 2010.

focuses on extreme weather–based supply shocks; however, this is a perfect example of a massive supply shock, but in this case it is politically oriented. As mentioned elsewhere, commodities have no biases. A supply shock is a supply shock whether it's from a politically driven move as with China or from a global climate shock. This is a key point to remember because either way commodity prices go up.

There are numerous other issues associated with China's being the absolute, dominant leader in rare earths, including the fear that certain producers in China may be disposing of by-product and radioactive uranium and thorium in environmentally unfriendly ways, but that is another story. The purpose of this segment is to assess the supply/demand situation. As is typically the case when prices rise, it stimulates interest for increased production. The United States, after not mining any rare earths in 2010, is feverishly working to tap into local sources to reduce dependence on China. So the potential does exist that in the future we could see the supply side ramp up significantly, but in the near to intermediate term (and possibly much longer, depending on the success of the ramp-up programs in the United States and other locations) the supply-side picture looks to remain quite tight.

So how do we play this opportunity? There are opportunities within the stock market and the ETF market. Within the stock market there are a handful of companies that are trying to ramp up mine production of rare earths. Unfortunately, most of these companies currently produce zero revenue. There is definitely a place for project-level financing and investing, but that is a riskier play than the average stock selection. There is a notable exception in the stock market. Molycorp Inc. (ticker MCP) is actually generating revenue. In fact, over the past 12 months its quarterly revenue is up by about 700 percent. This sounds terrific but its quarterly revenue most recently was only in the $25 million range—still quite small. Interestingly, despite the company's losing cash in its fledgling operation, its stock price is up by about 350 percent in the past 12 months. Clearly, the stock market is expecting great things from this company, as the stock price is already pricing in a very strong future.

Molycorp's mining assets are located in California. So, from the standpoint of an extreme weather–based investor, the two critical geographies to watch are Baotou City, and specifically the Daiyunebo Mine located in Inner Mongolia, and Mountain Pass, California, the location of the Molycorp project, which is fairly close to Las Vegas, Nevada. So in the event of a global climate shock impacting the specific region of Inner Mongolia, the impact would cause the price of rare earths to rise. This would be a home run for the ETF market as well as for Molycorp because it would be the last man standing that would directly benefit from rising rare earth prices. However, in the event that the Mountain Pass, California, project

TABLE 4.33 Rear Earth Elements Action Plan

	Where's the Flood? Baotou City, Inner Mongolia	Where's the Flood? Mountain Pass, California
Biggest Winners (Ranked)	Rare Earth ETFs Molycorp Inc.	None
Biggest Losers (Ranked)	Local Chinese Miners	Molycorp Inc.

Source: Public filings.

experienced a severe weather event, Molycorp would obviously be the biggest loser but because it is so small, the ETF market and the Chinese source would be essentially unaffected.

Taking these considerations into account, our action plan for weather-based rare earth investing is shown in Table 4.33.

POTASH

As discussed in Chapter 1, potash falls into the Jackpot category. Now we are going to drill down a bit farther into the key global locations for the mine production of potash. Table 4.34 depicts the attractive geographic concentration.

TABLE 4.34 Global Potash Mine Production

Global Potash Mine Production	Percent Split
Canada	29%
Russia	20%
Belarus	15%
China	9%
Germany	9%
Israel	6%
Jordan	4%
United States	3%
Chile	2%
Brazil	1%
Spain	1%
United Kingdom	1%
Ukraine	0%
Total	**100%**

Source: USGA, 2010.

TABLE 4.35 Potash Producers

Fertilizer Produced	Mosaic	Potash	Agrium	Uralkali	Sinofert
Potash	31%	46%	25%	100%	25%
Phosphates	69%	28%	22%	0%	22%
Nitrogen based	0%	26%	53%	0%	34%
Other	0%	0%	0%	0%	19%

Source: Public filings.

As shown, potash source rock has the desired geographic concentration for extreme weather–based investing. The dominant geography is Canada, holding 28 percent of the global mining production, followed by Russia at 20 percent of the total and Belarus at 15 percent of the total. More specifically, the Canadian mines are centered in Saskatchewan, and the Russian mines are in Berezniki in the Perm Territory.

Now let's take a look at the key global publicly traded potash producers. We also are interested in their product mix. Potash is only one of three major categories of fertilizer. The three types are potash, phosphates, and nitrogen-based fertilizers. The product mix for all of the major players is shown in Table 4.35. The public equity ticker symbols for the potash producers are shown in Table 4.36.

As shown, there is quite a wide range, with 100 percent of Uralkali's revenue coming from the potash nutrient, compared with 24 percent for Sinofert (the largest fertilizer maker in China). As we drill down one step further, we discover the location of the primary potash mines for each player along with their revenue exposure to the potash nutrient, as shown in Table 4.37.

With Table 4.37 at our fingertips it becomes very easy to predict who will be the winners and who will be the losers in the event of a global climate event in any of the top four geographies in the world. For example, if

TABLE 4.36 Equity Tickers for Potash Producers

Company	Stock Ticker
Mosaic	MOS
Potash	POT
Agrium	AGI
Sinofert	297 HK
Uralkali	URKA RM

Source: Public filings.

TABLE 4.37 Potash Producers

Company	Potash Sales	Canada*	Russia*	Belarus	China
Mosaic	31%	X			
Potash	46%	X			
Sinofert	24%				X
Agrium	25%	X			
Uralkali	100%		X	X	

*Specifically, Saskatchewan, Canada, and Perm Territory in Russia.
Source: Public filings.

the Perm Territory in Russia gets hit, obviously Uralkali will be the biggest loser and Potash Corporation will be the biggest winner.

Taking into account all of the possibilities, we come up with Table 4.38, which is our action plan for extreme weather–based investing in the potash nutrient (also see Chapter 5, "Global Climate Shock Number Three: Farmland Droughts, Floods, and Frost," where we also talk about additional opportunities in the potash nutrient from a farmland perspective).

MINING EQUIPMENT MAKERS

When dealing with mining-based global climate shocks, not only do we have investment opportunities with the commodities that are directly involved with the extreme weather event, but we also have tangential winners that do not make any commodities. The manufacturers of the equipment that is used in the mine are also winners. The key publicly

TABLE 4.38 Potash Nutrient Action Plan

	Where's the Flood? Canada*	Where's the Flood? Russia*	Where's the Flood? Belarus
Biggest Winners (Ranked)	Uralkali (100%) Sinofert (24%)	Potash (46%) Mosaic (31%) Agrium (25%) Sinofert (24%)	Potash (46%) Mosaic (31%) Agrium (25%) Sinofert (24%)
Biggest Losers (Ranked)	Potash (46%) Mosaic (31%) Agrium (25%)	Uralkali (100%)	Uralkali (100%)

*Specifically, Saskatchewan, Canada, and Perm Territory in Russia.
Source: Public filings.

TABLE 4.39 Equity Tickers for Mining Equipment Makers

Company	Stock Ticker
Caterpillar	CAT
Deere & Co.	DE
Bucyrus	BUCY
Joy Global	JOYG

Source: Public filings.

traded mining equipment makers in the commodity space are listed in Table 4.39.

The revenue mix is important because it defines the key sectors for each equipment maker. The percent revenue split for each of these names is shown in Table 4.40.

As can be seen, in terms of revenue exposure to mining, Bucyrus and Joy Global are the clear winners, with 100 percent of their revenue derived from the mining space. To be more specific, we can split this up one step further between revenue derived from underground mining versus surface mining, as shown in Table 4.41.

As shown, both companies have exposure to underground as well as surface mining, but both have 100 percent of their revenue derived from mining as a general category.

Behind Joy Global and Bucyrus, we have Caterpillar holding approximately 12 percent of its total revenue from the mining sector. In last place we have Deere & Co., with zero percent of revenue from the mining space. However, in the case of Deere & Co, we can see that it has obvious exposure to the agricultural space, with 78 percent of its total revenue feeding this end market. So we will save our discussion on Deere & Co. for Chapter 5, "Global Climate Shock Number Three: Farmland Drought, Floods, and Frost."

TABLE 4.40 Mining Equipment Makers

% of Revenue	Caterpillar	Deere & Co.	Bucyrus	Joy Global
Agricultural	0%	78%	0%	0%
Mining	12%	0%	100%	100%
Earth moving	15%	0%	0%	0%
Other	73%	22%	0%	0%
Total	**100%**	**100%**	**100%**	**100%**

Source: Public filings, 2010.

TABLE 4.41 Mining Equipment Pure Plays

% of Revenue	Bucyrus	Joy Global
Underground Mining	35%	58%
Surface Mining	65%	42%
Total	**100%**	**100%**

Source: Public filings.

As we have discovered in this chapter, *mining* is a very general term because there are literally dozens of commodities associated with mining. It therefore becomes challenging to link to Bucyrus or Joy Global to specific extreme weather events. Nevertheless, it is a well-known fact that when the miners are doing well, so are the equipment makers.

Because we know that coal is a significant part of the end-market mix for these two equipment makers, the stock price of both Joy Global and Bucyrus was compared against the leading thermal coal maker in the United States, Peabody Coal (ticker BTU). A remarkably high correlation between Joy Global and Peabody stock prices was found, as shown in Figure 4.1.

The average correlation over the prior decade was generally in excess of 90 percent between these two stocks. Interestingly, there was also a

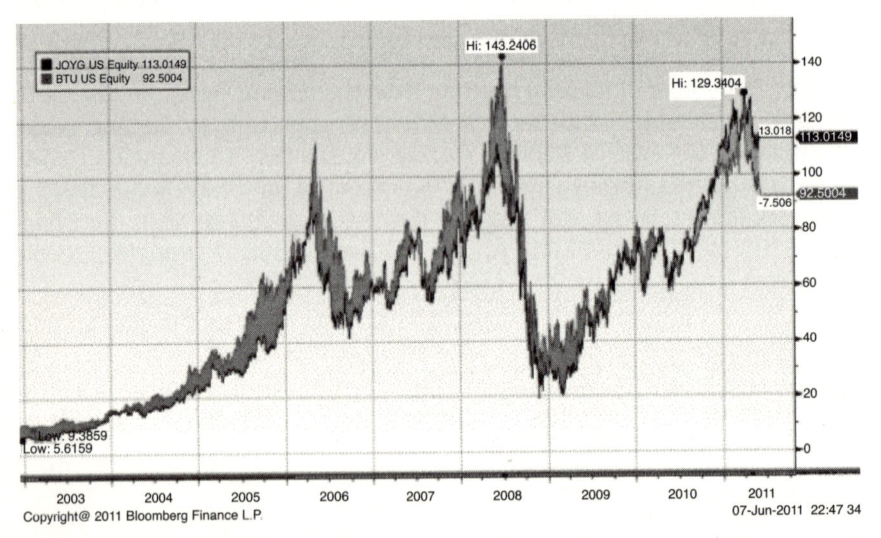

FIGURE 4.1 Peabody Coal (BTU) versus Joy Global (JOYG) Stock Price, 2002 to 2011
Source: Used with permission of Bloomberg Newswire Permissions.

very high correlation between both equipment makers and BHP, a leading globally diversified miner, as well as a strong correlation between Bucyrus and Peabody. The best correlation, however, was between Joy Global and Peabody Coal Company.

So how do we use this information on the mining equipment makers? There are a couple of key points here:

- Because the equipment makers are one step removed from the commodity supply shock, the specific commodity should be the investment of choice in preference over the equipment makers.
- Now that we know that there is a high correlation between these two primary mine equipment makers and the coal sector and the mining sector in general, we can potentially invest in the equipment makers as a diversification play, which still benefits from the favorable mining outlook.

CHAPTER 5

Global Climate Shock Number Three: Farmland Droughts, Floods, and Frost

Yes, even the farmers can cash in on global climate change—as long as it is not his farm that is seeing the drought or the floods! If it is some other region that is seeing the floods and the drought, we get the classic supply shock and agricultural commodity prices will skyrocket. *Of course, the more widespread the drought/flood region and the longer the duration, the more dramatic the commodity price spike reaction.*

Whether the farmer is growing corn, soybeans, wheat, cotton, coffee, cocoa, or any other farm product, there is an optimal level of water needed for maximum quantity per acre yields. Any extreme, whether it is a drought or a flood, will severely limit the harvest, thus creating a supply shock in the market that will result in prices going higher. Droughts in certain areas can be diminished with the use of irrigation, but extreme conditions of any kind are very damaging.

When analyzing the universe of agricultural commodities, it is apparent that when the price of corn or soybeans or wheat or any of the other agricultural commodities rise, the farmer reaps the profit. We cannot take advantage of this unless we own the farm. So, from an investor's standpoint, we need to be more creative to take advantage of these bull markets, which are highly prone to supply shocks associated with global climate change today and likely even more so in the future. Fortunately, there are numerous opportunities. The most direct method is the futures market, considering that the stock and bond market options are extremely limited because farms are privately owned by farmers. This method of investing is a very straightforward bet on the price of the commodity. It is simplistic in the sense that we don't care so much where or what causes the supply shock—we benefit regardless. For example, we recently saw the worst drought in Russia in more than half a century, extremely severe flooding in Australia, excessive rainfall in Canada, and drier-than-normal conditions in Europe, all of which drove up the price of the agricultural commodities. As will be shown in this chapter, the impact of the combination of good supply/demand fundamentals on top of frequent (and getting more frequent) supply shocks has a positive price impact on the key agricultural commodities, including corn, soybeans, wheat, cotton, coffee, cocoa, and many others. Interestingly, if we drill down and find out where each of the key agricultural commodities is produced globally, we can rank them in terms of their investing attractiveness depending on what region of the world is experiencing severe drought or flood conditions. First, we will talk about the location of production for each of the major agricultural commodities, and then we will tie this whole section together by creating an action plan and a specific ranking of the most attractive through to the least attractive futures market investment in a simplified table format.

Let's start with sugar.

SUGAR

As discussed in Chapter 1, sugar is in growing demand globally and the supply side is susceptible to the vagaries of the weather. Importantly, sugar also meets the investing criteria of geographic concentration, meaning a few areas of the world hold a large percentage of the total sugar production. As discussed earlier, this is critical because in the event that a high-percentage region is struck by severe, widespread drought, for example, this will represent a global supply shock. The sudden lack of availability of sugar will drive up sugar prices in the face of relentless demand for sugar. The global geographic breakdown for the production of raw sugar is shown in Table 5.1.

As can be seen in Table 5.1, Brazil and India are very important regions of the world for sugar production, particularly in the south central region of Brazil. Interestingly, within each country, there is often a highly concentrated region that houses much of the sugar production. For example, the Guanxi region of China houses approximately 65 percent of the sugar production, and if you add in the sugar produced in Yunnan, Guangdong, Hainan, and Xinjiang, the total comes to 95 percent for the Chinese region.

It is obvious from the table that if Brazil or India were to experience widespread drought or flooding conditions, it would represent a severe supply shock to the sugar market and the price of sugar in the futures market would rise. (See Chapter 9, "Playing Both Sides of the Coin," to see how to simultaneously bet that sugar prices will go up and that it may

TABLE 5.1 Global Raw Sugar Production

Raw Sugar Production	Tons in Millions	Percent Split
Brazil (south central)	39.4	24%
India	25.7	16%
EU-27	14.8	9%
China	12.7	8%
United States	7.6	5%
Thailand	6.9	4%
Mexico	5.5	3%
Australia	4.8	3%
Pakistan	3.3	2%
Russia	2.9	2%
Turkey	2.4	1%
Argentina	2.3	1%
Columbia	2.2	1%
South Africa	2.1	1%
Ukraine	2.0	1%
Egypt	1.8	1%
Canada	0.1	0%
Other	25.5	16%
World	**162**	**100%**

Source: USGS, 2010.

hurt the stock prices of buyers of sugar, including chocolate and soft drink producers.)

After we talk about the global production concentration for each of the major agricultural commodities, we will tie them all together toward the end of this chapter.

COFFEE

Similar to essentially all commodities, the demand for coffee is on the rise, as discussed in Chapter 1. In terms of production, coffee is particularly attractive because it also meets one of the key criteria of investing in extreme weather events. Specifically, it has geographic concentration of production. The key geographic regions of coffee bean production are shown in Table 5.2.

As shown in Table 5.2, when it comes to coffee bean production, once again, Brazil is the big winner. Also important in the production of coffee beans is Vietnam, Colombia, and Indonesia. If any one of these regions experienced severe weather in the form of droughts or floods, the price of coffee in the futures market would rise.

TABLE 5.2 Global Coffee Bean Production

Coffee Bean Production	60-kg Bags (in Millions)	Percent Split
Brazil	55	39%
Vietnam	19	13%
Colombia	9	6%
Indonesia	9	6%
India	5	4%
Mexico	5	3%
Ethiopia	4	3%
Guatemala	4	3%
Peru	4	3%
Honduras	4	3%
Other	22	17%
World	**139**	**100%**

Source: USDA, 2010.

We will tie the results for coffee together with the other agricultural commodities after we discuss them individually.

COTTON

Cotton is another exciting agricultural commodity. Not only is the demand side of the picture in full swing, but the supply side meets the key weather-based investing criteria of geographic concentration in production. The global geographic split in cotton production is shown in Table 5.3.

The key regions in the world for cotton production are China, India, the United States, Pakistan, and Brazil. Severe drought or flood damage in any of these cotton-producing regions would have an upward impact on

TABLE 5.3 Global Cotton Production

Cotton	Bales (480-Pound bales; in Millions)	Percent Split
China	33	29%
India	25	22%
USA	17	15%
Pakistan	11	9%
Brazil	7	6%
Other	22	19%
World	**114**	**100%**

Source: USDA, 2010.

the price of cotton. Interestingly, there are other supply shock effects in the cotton market. As shown, India is an enormous producer of cotton. When India decided to put a ban on exports of cotton, it had the same basic effect as a weather-based supply shock. This is true because other countries consume cotton that is produced in India, and when that source is no longer available, buyers from other countries get nervous and become willing to increase the price they will pay to get their cotton. A ban on exports from countries is not uncommon and is often driven by reasons of satisfying local, domestic demand first to make sure their own people get their needs met. Export bans can also occur for other political reasons, but regardless of the reason, it has the same basic, positive, and beneficial effect as a weather-based supply shock. No matter how you slice it, we win in the futures market under conditions of a severe supply shock regardless of the cause.

COCOA

Similar to the other commodities, global demand for cocoa is growing. As you know from Chapter 1, the key end-market demand driver for cocoa is chocolate, and everybody knows that chocolate is in high demand globally.

In addition, the supply side of cocoa is very attractive because it also meets our weather-based investing criteria of geographic concentration. Cocoa, however, is elite in terms of its much higher than average level of geographic concentration. Table 5.4 shows the geographic split for cocoa bean production.

As shown the key cocoa producing regions of the world are Cote d'Ivoire (Ivory Coast, a relatively small area on the coast of northwest

TABLE 5.4 Global Cocoa Bean Production

Cocoa Bean Production	Percent Split
Cote d'Ivoire	37%
Ghana	21%
Indonesia	13%
Cameroon	5%
Nigeria	5%
Brazil	5%
Ecuador	3%
Dominican Republic	1%
Other	10%
World	**100%**

Source: USDA, 2010.

Africa), Ghana, and Indonesia. Three additional second-tier producers are Cameroon, Nigeria, and Brazil. Any weather disruption to the Cote d'Ivoire region would obviously have a massive supply shock impact and correspondingly put upward pressure on the price of cocoa beans. This agricultural commodity is particularly attractive because the top geographic region is relatively small, thus making it more susceptible to widespread damage from extreme weather–based supply shocks. Another very attractive feature of geographic concentration in this region of the world is that it is subject to political supply shocks as well. See Chapter 8, "Real-Life Examples: Execution, Results, and Timing," to see how cocoa bean prices responded recently to a real-world supply shock.

CORN

The global demand for corn is growing (see Chapter 1), and the supply is also limited. Corn has a great story, as discussed previously. When looking at corn, we see that it meets our weather-based investment criterion of geographic concentration. Similar to cocoa beans, corn is elite in terms of its much higher than average geographic concentration of production. The global split out on corn production is shown in Table 5.5.

As shown, the biggest producer of corn, by far, is the Corn Belt of the United States. Other individually big producing countries include China and Brazil. The EU-27 is a collection of 27 European countries and

TABLE 5.5 Global Corn Production

Corn Production	Percent Split
United States	41%
China	19%
EU-27	7%
Brazil	7%
Argentina	3%
Mexico	3%
India	2%
South Africa	2%
Ukraine	1%
Canada	1%
Other	14%
World	**100%**

Source: USDA, 2010.

therefore has less geographic concentration than we like to see, but that is fine because we have plenty of concentration within the United States.

We will see how corn stacks up against the other agricultural commodities after each commodity is reviewed individually.

SOYBEANS

Soybeans also fall into the Jackpot category, as mentioned previously. The demand is solid globally, and the supply side, like corn, has very favorable geographic concentrations. The production of soybeans globally is shown in Table 5.6.

As shown, in the event of widespread drought or flooding in the Corn Belt region of the United States, Brazil, or Argentina, a significant supply shock in the availability of soybeans would result.

Again, we will see how soybeans compare to other agricultural commodities toward the end of this chapter, after we discuss each agricultural commodity individually.

WHEAT

Although the supply/demand fundamentals for wheat are favorable, as discussed earlier, it has medium-level geographic concentration of production. It is not nearly as attractive as corn and soybeans from the standpoint of our key weather-based investing criterion of geographic concentration, but nevertheless still offers us some opportunities. The geographic splitout on wheat production globally is shown in Table 5.7.

TABLE 5.6 Global Soybean Production

Soybean Production	Percent Split
United States	37%
Brazil	25%
Argentina	19%
China	7%
India	4%
Other	8%
World	**100%**

Source: USDA, 2010.

TABLE 5.7 Global Wheat Production

Wheat Production	Metric Tons (in Millions)	Percent Split
China	112	16%
India	79	11%
United States	68	10%
Russia	64	9%
France	39	6%
Canada	29	4%
Germany	26	4%
Ukraine	26	4%
Australia	21	3%
Pakistan	21	3%
Other	205	30%
World	**690**	**100%**

Source: USDA, 2010.

As shown, although wheat is not as attractive as corn or soybeans, it still has some degree of geographic concentration, and therefore can benefit from supply shocks in the event severe drought hits any of the key regions, including China, India, the United States, or Russia. In fact, last year, Russia suffered from its worst drought in more than 50 years. We explore this drought and its impact on wheat prices in Chapter 8.

ORANGES

Oranges are a wonderful commodity given their growing demand and concentrated global geographic supply. See Chapter 1 for additional details on the favorable supply/demand dynamics behind the orange crop. Looking globally into the production of oranges, we arrive at Table 5.8.

As shown, Brazil and the United States account for nearly half of the world's orange crop. This is very, very favorable from the standpoint of our key investing criterion of global geographic concentration. Even more impressive is that within the United States orange production is split between Florida and California, with the former holding 80 percent of total U.S. production and the latter holding the remaining 20 percent. Within Brazil, Sao Paulo holds the lion's share of orange production. This means that in the event of drought or flood or even the occasional frost (oranges hate frost) in Florida, the orange juice price will rise. This will hold true if there is a global climate shock in any of the key regions, including Florida, California, and Sao Paulo.

TABLE 5.8 Global Orange Production

Orange Production	**Percent Split**
Brazil (mostly Sao Paulo)	27%
United States (80% Florida/20% California)	13%
Mexico	6%
India	6%
China	5%
Spain	5%
Italy	4%
Other	34%
World	**100%**

Source: USDA, 2010.

We will see how oranges stack up against other agricultural commodities in the next section.

TYING TOGETHER ALL OF THE AGRICULTURAL COMMODITIES

Now is the moment we have all been waiting for. Which of the eight agricultural commodity crops that we just reviewed is the best in which to invest your money? Well, when it comes to investing based on global climate shocks, the answer is that it depends on the geographic region of the world that is seeing the drought, flood, or frost. Let's summarize the findings for all eight of these agricultural commodities into a single table, showing their respective geographic concentrations (Table 5.9).

When looking at this table, what we are really interested in is the high percentage figures. We want to identify the concentrated geographies that make markets move. We will call these key geographies the Global Agricultural Commodity "A" Team. The "A" Team consists of the 10 global regions of the world shown in Table 5.10.

When any one of these geographies (or more specifically the regions discussed earlier within each of these geographies) experiences a global climate shock in the form of severe drought, severe flooding, or severe frosting (frosting mostly for oranges), commodity prices will move upward—but which commodities? The tables that follow identify the key commodities for each Global "A" Team region of the world. Not only do the tables identify the key commodities, but they rank them in investing attractiveness! With this list at your fingertips, you are in a far better position than the rest of the investing universe to take quick action when a

TABLE 5.9 Global Geographic Concentration of Major Agricultural Commodities

Country	Sugar	Coffee	Cocoa	Oranges	Wheat	Soybeans	Corn	Cotton
Argentina	1%					19%	3%	
Australia	3%				3%			
Brazil	24%	39%	5%	27%		25%	7%	6%
Canada	0%				4%		1%	
China	8%			5%	16%	7%	19%	29%
Columbia	1%	6%						
Egypt	1%							
EU-27	9%						7%	
India	16%	4%		6%	11%	4%	2%	22%
Mexico	3%			6%			3%	
Other	16%					8%	14%	19%
Pakistan	2%							9%
Russia	2%				9%			
South Africa	1%						2%	
Thailand	4%							
Turkey	1%							
Ukraine	1%						1%	
United States	5%			13%	10%	37%	41%	15%
Ethiopia		3%						
Guatemala		3%						
Honduras		3%						
Indonesia		6%	13%					
Mexico		3%						
Peru		3%						
Vietnam		13%						
Cameroon			5%					
Cote d'Ivoire			37%					
Dominican Republic			1%					
Ecuador			3%					
Ghana			21%					
Nigeria			5%					
France					6%			
Germany					4%			
Pakistan					3%			
Ukraine					4%			
Italy				4%				
Spain				5%				
Other		16%	10%	33%	30%			
Total	**100%**	**100%**	**100%**	**100%**	**100%**	**100%**	**100%**	**100%**

Source: USDA, 2010.

TABLE 5.10 Global "A" Team

The Global "A" Team
Brazil
Argentina
China
India
Russia
United States
Indonesia
Vietnam
Cote d'Ivoire
Ghana

Source: USDA, 2010.

particular region gets hit with a global climate shock. We will start with Brazil as shown in Table 5.11.

So how do we interpret Table 5.11? As you can see, the key question in this table is "Where is the farmland drought/flood/frost?" The answer in this particular example is Brazil. If that is indeed the case, then the biggest potential winners are coffee, oranges, soybeans, and sugar—in that order! The agricultural commodities that will still benefit but not to the same extent are corn, cotton, and cocoa. The reason they will not benefit as much is that (as you can see in the parentheses next to each commodity) their exposure to the Brazilian region is much lower. Specifically in the case of coffee, 39 percent of global coffee production is in Brazil. So,

TABLE 5.11 Brazil Agricultural Commodity Action Plan

	Where's the Farmland Drought/ Flood/Frost? Brazil
Biggest Winners (Ranked)	Coffee (39%)
	Oranges (27%) (particularly if in Sao Paulo)
	Soybeans (25%)
	Sugar (25%)
	Corn (7%)
	Cotton (6%)
	Cocoa (5%)
Biggest Losers	The farmers whose fields were impacted

Source: USDA, 2010.

obviously, if that particular region is suffering from a widespread global climate shock, then coffee will feel the greatest impact on the availability of supply and therefore the greatest potential increase in the price in the futures market.

On the other side of the coin, the only direct loser in this analysis is the farmland whose acreage was directly impacted by the global climate shock. Now, there are actually some secondary losers in this analysis as well. Take, for example, the *people* in the United States who enjoy their morning coffee. They may be forced to spend a little extra per cup because of this global climate shock.

In addition, some *manufacturers* will also see a secondary negative/loser effect. For example, when the price of sugar goes up, companies like Coca-Cola and Hershey, which buy a large amount of sugar, will feel the pain of this global climate shock. Similarly, when the cost of coffee beans skyrockets, Sara Lee feels the pain. For more information on how to play the winners and losers simultaneously, see Chapter 9, "Playing Both Sides of the Coin."

See Chapter 8 for detailed real-life examples and their results and how to use this information during future extreme weather events. In addition, refer to Chapter 8 to see the list of extreme weather based investing rules.

The second country in the Global "A" Team is Argentina. Table 5.12 shows our action plan in the event that Argentina is hit by a widespread drought or flood.

The results in Table 5.12 are not quite as exciting because Argentina houses far fewer agricultural crops as compared to Brazil. Nevertheless, in the event that Argentina gets hit by widespread drought or floods, the soybean market would be the biggest winner because Argentina represents 19 percent of global production of soybeans. Corn and sugar would be winners but on a much smaller scale due to the relatively small exposure to this geography.

The third country in the Global "A" Team is China. Table 5.13 shows our action plan in the event China is hit by widespread drought or flood, particularly in the region that produces each crop listed.

TABLE 5.12 Argentina Action Plan

	Where's the Farmland Drought/Flood? Argentina
Biggest Winners (Ranked)	Soybeans (19%) Corn (3%) Sugar (1%)
Biggest Losers	The farmers whose fields were impacted

Source: USDA, 2010.

TABLE 5.13 China's Agricultural Commodity Action Plan

	Where's the Farmland Drought/Flood? China
Biggest Winners (Ranked)	Cotton (29%) Corn (19%) Wheat (16%) Sugar (8%) Soybeans (7%) Oranges (5%)
Biggest Losers	The farmers whose fields were impacted

Source: USDA, 2010.

As shown in the table, cotton, corn, and wheat are the big crops in China, with cotton the most important because 29 percent of global cotton is produced in China and therefore any sizable negative weather event in the cotton-producing regions of China will very likely impact the market and drive cotton prices higher. To a lesser extent, sugar, soybeans, and oranges would also be big winners for this region.

The fourth country in the Global "A" Team is India. Table 5.14 shows our action plan in the event that India is hit by a widespread drought or flood, particularly within the regions of India producing these major crops.

As shown in Table 5.14, the three biggest winners in this region are cotton, sugar, and wheat—in that order! Also benefiting from an India-based global climate shock are oranges, coffee, soybeans, and corn (in that order) but to a lesser extent than the top three. Again, this is simply due to the higher level of exposure that cotton, sugar, and wheat have to India.

TABLE 5.14 India's Agricultural Commodity Action Plan

	Where's the Farmland Drought/Flood? India
Biggest Winners (Ranked)	Cotton (22%) Sugar (16%) Wheat (11%) Orange Juice (6%) Coffee (4%) Soybeans (4%) Corn (2%)
Biggest Losers	The farmers whose fields were impacted

Source: USDA, 2010.

TABLE 5.15 Russian Agricultural Commodity Action Plan

	Where's the Farmland Drought/Flood? Russia
Biggest Winners (Ranked)	Wheat (9%)
	Sugar (2%)
Biggest Losers	The farmers whose fields were impacted

Source: USDA, 2010.

The fifth region in the Global "A" Team is Russia. Table 5.15 shows our action plan in the event that Russia is hit by a widespread drought or flood, particularly within the regions of Russia producing these major crops.

As shown Table 5.15, Russia is generally lacking as a major leader within the agricultural commodity space, but nevertheless 9 percent of global wheat is made in Russia. In the event of a severe widespread drought or flood, the wheat price would and has in the past responded with a price move upward. After all, if 9 percent of an already-tight global commodity disappears overnight, it will most certainly frighten the buyers, and the price of wheat will go higher. Last year, in fact, Russia suffered the worst drought in more than five decades. We cover the details of this drought in Chapter 8.

The sixth region in the Global "A" Team is the United States. Table 5.16 shows our action plan in the event that the United States is hit by a widespread drought or flood, particularly within the regions producing these major crops.

As shown in Table 5.16, the United States is quite impressive in terms of its very consolidated geographic exposure particularly to corn, soybeans, cotton, oranges, and wheat—in that order! The last "biggest winner," sugar, pales by comparison, but is a winner nonetheless. Interestingly, the biggest

TABLE 5.16 U.S. Agricultural Commodity Action Plan

	Where's the Farmland Drought/Flood? United States
Biggest Winners (Ranked)	Corn (41%)
	Soybeans (37%)
	Cotton (15%)
	Oranges (13%)
	Wheat (10%)
	Sugar (5%)
Biggest Losers	The farmers whose fields were impacted

Source: USDA, 2010.

TABLE 5.17 Indonesia Agricultural Commodity Action Plan

	Where's the Farmland Drought/Flood? Indonesia
Biggest Winners (Ranked)	Cocoa (13%) Coffee (6%)
Biggest Losers	The farmers whose fields were impacted

Source: USDA, 2010.

winner, corn, also lends itself to many additional secondary investing opportunities within the stock market. These secondary investments are covered in several areas in detail in various other chapters.

The seventh region in the Global "A" Team is Indonesia. Table 5.17 shows our action plan in the event that Indonesia is hit by a widespread drought or flood, particularly within the regions producing these major crops.

Although Indonesia is not as exciting as the United States, two major crops still show up. Both cocoa and coffee are the big winners in this sector as this region holds enough of these commodities that if their supply availability was interrupted due to a global climate shock, it would likely move the market and cause the price of cocoa and coffee to rise. Again, as a reminder, see Chapter 13, "Basic Principles of Futures Market Investing," for the mechanics on how to take advantage of these global agricultural commodity plays that we are covering in this chapter.

The eighth region in the Global "A" Team is Vietnam. Table 5.18 shows our action plan in the event that Vietnam is hit by a widespread drought or flood, particularly within the regions producing these major crops.

When looking at Table 5.18, it is clear that the Vietnam story is quite simplistic with only a single agricultural commodity of significance in the region. Basically, in the event of a global climate shock in Vietnam, the commodity to watch is coffee pricing, as they make a meaningful amount of the globally produced coffee, at 13 percent of the total.

The ninth region in the Global "A" Team is Cote d'Ivoire. Table 5.19 shows our action plan in the event that this region is hit by a widespread

TABLE 5.18 Vietnam Agricultural Commodity Action Plan

	Where's the Farmland Drought/Flood? Vietnam
Biggest Winners	Coffee (13%)
Biggest Losers	The farmers whose fields were impacted

Source: USDA, 2010.

TABLE 5.19 Cote d' Ivoire Agricultural Commodity Action Plan

	Where's the Farmland Drought/Flood? Cote d'Ivoire
Biggest Winners	Cocoa (37%)
Biggest Losers	The farmers whose fields were impacted

Source: USDA 2010

drought or flood, particularly within the regions producing these major crops.

Table 5.19 is particularly attractive in two respects. Not only is it simple in that this region has only a single agricultural commodity of significance, but it is extraordinarily, geographically concentrated. In the event of a global climate shock in this region, the obvious and very large and positive impact would come to the cocoa bean as the price of cocoa beans would rise.

The tenth and final Global "A" Team region is Ghana. Ghana holds nearly the same attractive story as with Cote d'Ivoire. Again, it is simplistic in its unilateral focus on cocoa and attractive in the sense that it is a highly concentrated region in that 21 percent of the world's cocoa is made in Ghana as shown in Table 5.20. Therefore, in the event of a major global climate shock in Ghana, the price of cocoa would increase, thus providing major opportunities for investors of cocoa in the futures market.

Similar to the discussion on sugar, when the price of cocoa beans goes up significantly, this has a negative effect on buyers of cocoa beans, such as the Hershey Company. See Chapter 9, where we talk about how to make money on both the winners and losers simultaneously.

Another excellent way to make money on these agricultural commodities is through the second-tier effects. Specifically, we can focus on the stock price (or on the corporate bonds) of the companies that supply products to the farmer, including fertilizers, seeds, and agricultural chemicals. In addition, we will cover equipment makers in the agricultural space.

TABLE 5.20 Ghana Agricultural Commodity Action Plan

	Where's the Farmland Drought/Flood? Ghana
Biggest Winners	Cocoa (21%)
Biggest Losers	The farmers whose fields were impacted

Source: USDA, 2010.

We will look in detail at each one of these investing paths starting with fertilizers.

FERTILIZERS

As the price of corn goes higher and higher after a global climate shock to a key corn-producing region of the world, the farmer has incentive to increase his bushel-per-acre yield of corn because the number of acres he has available to him is limited. One way he can achieve this is by adding fertilizer to the field. This is a key demand driver for fertilizers. On the pricing side of the fertilizer market, corn prices are related to fertilizer prices. When corn prices are strong, the farmer has an incentive to plant more acres and to buy more fertilizer in order to increase the per-acre yield of corn. The increased demand for fertilizer helps to push up the price of these valuable fertilizer nutrients. Figure 5.1 shows the price of corn compared with the price of the three key fertilizer types, including North American potash, solid phosphates, and urea over the past seven years on a quarterly basis.

As shown, there is an obvious visual correlation between the price of corn and the price of fertilizer. Upon closer examination of the graph, we can see that the price of corn actually predicts the price of fertilizer as a

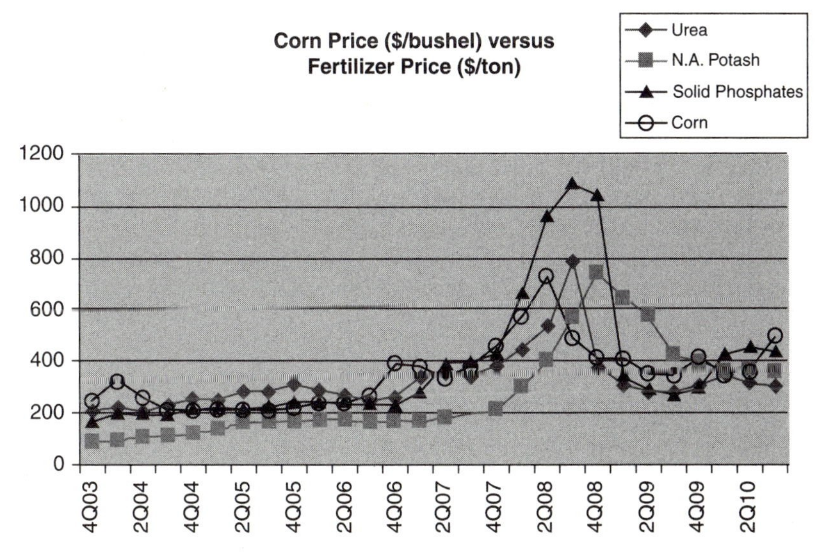

FIGURE 5.1 Corn Price versus Fertilizer Price
Source: Public company filings.

TABLE 5.21 Fertilizer Producers

Fertilizer Produced	Mosaic	Potash	Agrium	CF	Yara	Uralkali
Potash	31%	46%	25%	0%	0%	100%
Phosphates	69%	28%	22%	20%	0%	0%
Nitrogen based	0%	26%	53%	80%	100%	0%

Source: Public filings.

leading indicator. In fact, the correlation between the mineral potash and corn is +54 percent as shown in Figure 5.1, but it improves to +88 percent if we compare current corn prices to the price of potash two quarters into the future. Interestingly, for both urea and solid phosphates, the ability of corn to predict future pricing of these two materials is best only one quarter into the future. The data show that the corn-to-urea correlation is +66 percent but climbs to +86 percent when looking one quarter into the future. Similarly, the correlation for solid phosphates improves from +74 percent to +87 percent when looking one quarter into the future. If these relationships continue to hold true, then the recent spike in the price of corn to the current $7/bushel will have a beneficial pricing effect on the fertilizer market and will continue to have a beneficial effect as supply shocks continue to plague an already-tight global corn market. The direct connection between the price of corn and the price of fertilizer and hence the margins and profits of fertilizer producers is the needed connection that ensures that fertilizer makers benefit from agricultural commodity, weather-related supply shocks. As shown in Figure 5.1 and the preceding discussion, all three of the major fertilizer types benefit from increasing crop prices.

When looking at the publicly traded names in the fertilizer sector, we come up with Table 5.21. The equity ticker symbols for the fertilizer producers are shown in Table 5.22.

TABLE 5.22 Equity Tickers for Fertilizer Producers

Company	Stock Ticker
Mosaic	MOS
Potash Corp.	POT
Agrium	AGU
CF	CF
Yara	YAR No
Uralkali	URKA Rm

Source: Public filings.

Within North America, there are four major players, including Mosaic, Potash, Agrium, and CF Industries. They are all generally diversified across the various types of fertilizer, but each has a specialty, including phosphates for Mosaic, potash (the nutrient) for Potash Corporation, nitrogen-based and downstream retail operations for Agrium, and nitrogen-based for CF Industries. Internationally, we have the nitrogen-based company Yara based in Norway, and the Russian potash pure play Uralkali. There are other makers of fertilizer, but the names listed have a very high level of fertilizer concentration in their total revenue mix, and therefore they are attractive to us as extreme weather–based investors. It is good to keep a tab on which companies focus on which nutrients, but the good news here is that the price and margins of all three primary nutrients correlate highly with the price of corn. So, in general, the "whole space trades together" (meaning that the stock prices for all of the fertilizer makers tend to track each other) and typically upward as increased pressure hits the corn market in terms of tight stocks-to-use ratio (i.e., low inventory and high demand for corn) and the chronic onset of weather-induced supply shocks. So in other words when a global supply shock hits the corn market, a secondary way to make money is to buy stocks (or bonds) of the fertilizer makers.

In addition to the excitement in the fertilizer sector associated with rising corn prices, the sector is also very hot because each of these companies has been or will possibly be again a target of an acquisition, which always puts upward price pressure on a stock.

After we talk about the companies that produce seeds and agricultural chemicals in the next section, we will tie them all together and see how they stack up against the fertilizer producers in their investing attractiveness.

SEEDS AND AGRICULTURAL CHEMICALS

The seed and agricultural chemical guys also make out on the deal when floods and droughts destroy farmland products and consequently drive up the price of corn and other key agricultural commodities Similar to the discussion on fertilizers, the farmer has a perpetual desire to increase his per-acre yield of the grain he is producing whether it is corn, soybeans, wheat, cotton, or any of the other agricultural products. He accomplishes this in many ways and, in fact, regularly uses all of these methods simultaneously.

The methods available to the farmer include using fertilizers, as discussed earlier. He can also get much, much more creative by using genetically modified seeds with specific traits that allow them to actually be more resistant to insects, weeds, droughts, and many other specific crop ailments on a crop-by-crop basis. Another method that the farmer regularly

TABLE 5.23 Agricultural Seed and Chemical Producers

% of Total Revenue	Monsanto	DuPont	FMC	Chemtura	Dow	Syngenta	BASF
Seed	72%	29%	0%	0%	5%	24%	3%
Ag Chemical	28%	0%	40%	13%	5%	76%	3%
Other	0%	71%	60%	87%	90%	0%	94%
Total	**100%**	**100%**	**100%**	**100%**	**100%**	**100%**	**100%**

Source: Public filings.

uses is the chemical application method. In this case, he applies herbicides to kill weeds, insecticides to kill numerous kinds of bugs, and other types of chemicals to deal with other microorganisms.

The key players in the agricultural seed and agricultural chemical markets are shown in Table 5.23. Notice that in this case we are really not as concerned about the geography of these players because droughts or floods on farmland anywhere in the world will continue to benefit all of these producers because this global supply shock will increase the price of the grain. When the farmer makes more profits on his grain, his appetite to buy fertilizers, special seeds, and agricultural chemicals increases greatly.

Another way to present this is to combine seeds and agricultural chemicals into a single group, which is the more relevant way of thinking of it, considering that both seeds and agricultural chemicals benefit from grain supply shocks. This is shown in Table 5.24. The public equity ticker symbols for the seed and agricultural chemical makers are shown in Table 5.25.

As shown in Table 5.24, when adding together the seeds and agricultural chemical businesses, Monsanto and Syngenta are obviously the biggest winners because 100 percent of their revenue is in these two sectors. This is followed in rank order by FMC, with 40 percent of its revenue in these sectors; DuPont, with 29 percent of its revenue in these sectors; Chemtura, with 13 percent; Dow, with 10 percent; and BASF, with 6 percent of its revenue in these two sectors. In the next section we compare the investing attractiveness of all of the agricultural-based sectors covered in this chapter, including raw grain, fertilizers, seeds, and agricultural chemicals.

TABLE 5.24 Agricultural Seed and Chemical Producers

% of Total Revenue	Monsanto	DuPont	FMC	Chemtura	Dow	Syngenta	BASF
Seeds Plus Ag Chemicals	100%	29%	40%	13%	10%	100%	6%
Other	0%	71%	60%	87%	90%	0%	94%
Total	**100%**	**100%**	**100%**	**100%**	**100%**	**100%**	**100%**

Source: Public filings.

TABLE 5.25 Equity Tickers for Agricultural Seed and Chemical Makers

Company	Stock Ticker
Monsanto	MON
DuPont	DD
FMC	FMC
Dow	DOW
BASF	BASFY
Syngenta	SYT
Chemtura	CHMT

Source: Public filings.

Fertilizers vs. Seeds vs. Chemicals vs. Grains

If we combine all of the farmland-based investment opportunities associated with global climate shocks in the agricultural market, including the impact on fertilizers, agricultural seeds, agricultural chemicals, and the raw price of commodity grains into one single action plan, we arrive at Table 5.26. This table summarizes our action plan when widespread farmland sees massive droughts or floods (regardless of geography as discussed earlier).

TABLE 5.26 Action Plan Table for Widespread Farmland Droughts/Floods

	Farmland Droughts/Floods
Biggest Winners (Ranked)	Commodity Grain Futures Markets (100%)
	Monsanto (100%)
	Syngenta (100%)
	Potash Corp. (100%)
	Mosaic (100%)
	Agrium (100%)
	CF Industries (100%)
	Yara (100%)
	Uralkali (100%)
	FMC (40%)
	DuPont (29%)
	Chemtura (13%)
	Dow (10%)
	BASF (6%)
Biggest Losers	The farmlands seeing floods/droughts.

Note: Percentage of total revenue in the fertilizer OR seed OR Ag chemical markets.
Source: Public filings.

TABLE 5.27 Farmland and Mining Equipment Makers

% of Revenue	Caterpillar	Deere	CNH	Bucyrus	Agco	Joy Global
Agricultural	0%	78%	72%	0%	100%	0%
Mining	12%	0%	0%	100%	0%	100%
Earth Moving	15%	0%	18%	0%	0%	0%
Other	73%	22%	10%	0%	0%	0%
Total	**100%**	**100%**	**100%**	**100%**	**100%**	**100%**

Source: Public filings.

FARMLAND EQUIPMENT MAKERS

As we discussed in Chapter 4, equipment makers were a strong beneficiary to the positive fundamentals in the mining space. Likewise, equipment makers also benefit from the positive fundamentals in the agricultural space. The key publicly traded equipment makers and their associated revenue mix is shown in Table 5.27. The equity ticker symbols for the farmland and mining equipment makers are shown in Table 5.28.

In Chapter 4, covering flooding mines, we talked in detail about Bucyrus and Joy Global. In this chapter covering agricultural equipment, we can see Deere & Co, CNH, and Agco are the obvious biggest winners with very high percentages of their equipment revenue feeding this end market. Similar to Chapter 4, where we showed that both Joy Global and Bucyrus had a strong stock price correlation with Peabody Coal (the largest coal maker in the United States), in this chapter we show that Deere has a strong stock price correlation with leading fertilizer producer, Potash Corporation of Saskatchewan (ticker POT) in Figure 5.2. In addition, Agco and CNH Global have remarkably similar stock price movements over the past decade, as compared with both Deere and Potash Corporation.

TABLE 5.28 Equity Tickers for Equipment Makers

Company	Stock Ticker
Caterpillar	CAT
Deere & Co.	DE
Bucyrus	BUCY
Joy Global	JOYG
CNH Global	CNH
Agco Corp.	AGCO

Source: Public Filings

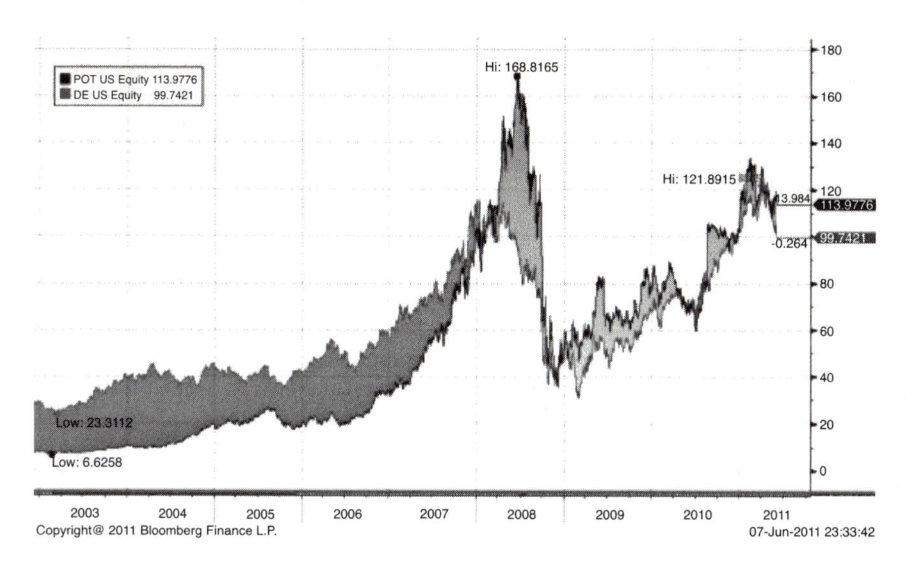

FIGURE 5.2 Deere & Co Stock versus Potash Corp. Stock, 2002–2011
Source: Used with permission of Bloomberg Newswire Permissions. Copyright © 2011. All rights reserved.

So how do we use this information?

- Similar to the equipment result from Chapter 4, the Deere & Co., Agco Global, and the CNH Global stocks should be used as a diversification play that enjoys the good fundamentals of the agricultural market.
- Also, in the event of a global climate shock or extreme weather event, the first investment choice should remain with the directly impacted commodity, while an investment in these highly correlated equipment makers should be a second-level-type investment.

CHAPTER 6

Global Climate Shock Number Four: Hurricanes and Tornadoes

Hurricanes and tornadoes are among the most severe weather patterns that can and will continue to emerge. These patterns obviously cause major and very often widespread damage, as we have experienced numerous times in the past. When analyzing the impact of extreme hurricanes, we learn that the beneficiaries to this event include North American independent natural gas/oil producers, housing construction, building materials, and infrastructure redevelopment. We will talk about each one of these in detail below.

INDEPENDENT NATURAL GAS PRODUCERS

Remember back in Chapter 1 we labeled natural gas as a Big Problem commodity due to an excessive supply of natural gas in North America. Recall also that we generally prefer to invest in the Jackpot commodities because the supply/demand picture is already tight even before the extreme weather event hits the market and therefore the relevant commodity price will see an even more extreme spike. However, despite the fact that natural gas currently falls into the Big Problem category, it will still respond favorably to supply shocks (particularly severe and long-term supply shocks) and therefore offers us an opportunity for making money, albeit possibly not quite as much as we would make in a Jackpot commodity under similar circumstances.

So the question we seek to answer is who are the winning natural gas producers when a hurricane strikes? In order to figure this out, let's start out with the global picture for natural gas. Table 6.1 shows the top-producing regions of natural gas around the world.

As shown in Table 6.1, both the United States and Russia hold a very material portion of the global production of natural gas. One would think that if Russia experienced an extreme weather event, the logical winner would be the U.S. producers. Unfortunately, this is generally not true because it is very difficult to export natural gas around the world. In order to export natural gas, it first needs to be cryogenically frozen into a liquid state at very, very cold temperatures. Now, there are liquefied natural gas (LNG) facilities that do, in fact, exist to achieve these cold temperatures for exporting natural gas, but the fact is that today they are not as prevalent as they need to be to make natural gas a fluently, globally traded commodity like its energy sister, oil. Natural gas is much more of a regionally priced commodity. Evidence that natural gas is regionally priced can be seen when comparing natural gas from the Middle East at 75 cents per million British thermal units (BTUs) with U.S. natural gas at $4 per million BTUs. Generally speaking, the high-cost players set the price of a globally priced commodity, but that is not occurring with the very cheap Middle Eastern natural gas. So what are we to do with natural gas, as extreme weather–based investors, when it doesn't follow the model of global pricing? The answer is that we look for regional opportunities.

Considering that the Gulf Coast region of the United States is flush with hurricanes every year, it seems like a very logical place to start. Let's

TABLE 6.1 Global Natural Gas Production

World Natural Gas Production	Percent
United States	20%
Russia	19%
Canada	5%
Iran	4%
Norway	3%
Qatar	3%
China	3%
Algeria	3%
Netherlands	3%
Saudi Arabia	3%
Other	34%
World Total	**100%**

Source: eia.gov, 2009.

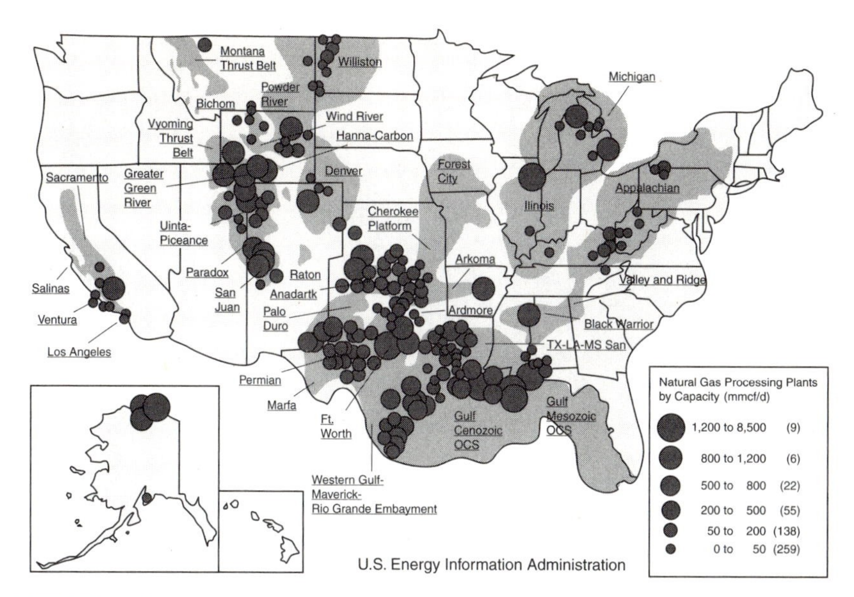

FIGURE 6.1 U.S. Natural Gas Processing Facilities
Source: EIA, 2010.

dig down and find out where natural gas is produced. Figure 6.1 shows the natural gas processing facilities around the country. These facilities are generally placed in close proximity to the original production of natural gas, and therefore give us a representative picture of where natural gas is produced in this country.

Amazingly, Texas and Louisiana account for nearly half of the U.S. processing capacity, with the largest amount along the Gulf Coast, directly in line with hurricane alley!

To be a little clearer on what is meant by the term *hurricane alley*, see Figure 6.2, which depicts the highest-probability locations for major weather events, including hurricanes, tornadoes, and, yes, even earthquakes within the U.S. borders.

As shown in Figure 6.2, the region of highest prevalence in hurricanes generally corresponds with the location of the production and processing of natural gas. This is a classic example of an extreme weather based investing opportunity.

Now that we've narrowed down the global opportunity to the United States, which companies should we focus on? As extreme weather–based investors, we like to get the biggest bang for our investing buck, so what we really are trying to find is the list of companies that have the greatest percentage of their revenue associated with exploration and production of natural gas and oil. (Technically speaking, most of the relevant companies

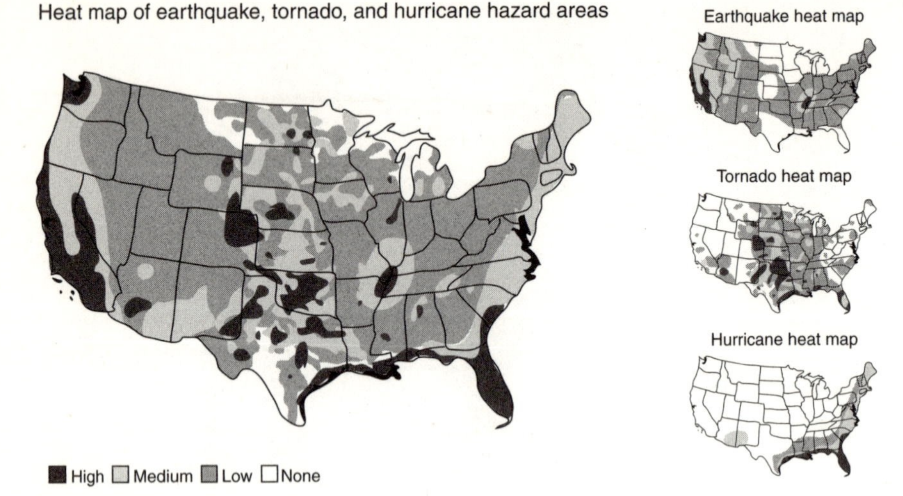

FIGURE 6.2 Map Showing Locations of Major U.S. Storms
Source: Underlying data from the Department of Commerce (NOAA); map courtesy of Pingdom AB.

of interest are also major explorers and producers of oil as well. These companies are migrating more and more toward drilling for liquids like oil as compared with natural gas because, as discussed earlier, oil is currently a Jackpot commodity and natural gas is not.) So does that mean we want a giant, integrated energy company like Exxon or a much smaller company that is focused strictly on finding natural gas and oil but doesn't own any downstream gas stations? To help you make the right choice, take a look at the value chain flowchart in Figure 6.3.

Monster-size companies like Exxon do all three steps of the value chain. As extreme weather–based investors, do we want that kind of company, or do we want a company that specializes in only one level of the value chain, and which level would that be? To answer the question, let's briefly talk about each box. The companies in the Upstream box sell natural gas and oil. When the price of these commodities spike upward, these companies win and win big.

The companies in the second box, Midstream, can also do quite well, but they are not as directly leveraged in general to the price of natural gas and oil. So as an extreme weather–based investor, so far, the Upstream box wins.

In the third and final box, Downstream, these companies actually buy crude oil and natural gas, so when the price of these commodities is skyrocketing, the Downstream companies can actually get squeezed. Not exactly a top choice for us. So, basically, to get the biggest bang for our

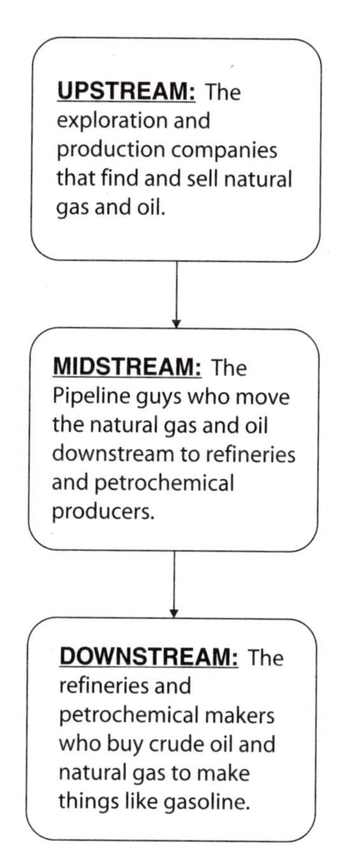

FIGURE 6.3 Energy Production Value Chain

investing buck as extreme weather–based investors, our focus is on box number one, Upstream.

Now that we know that the desired, target companies for us are in the Upstream category, we are one step closer to finding out which companies benefit the most during a Gulf Coast–based hurricane.

Using the same methodology we have used numerous times before, we already know that the "biggest loser" in this scenario has the highest percentage of their production located in hurricane alley. Granted, a hurricane has a mind of its own. It is essentially impossible to predict with precision where it will hit, so as extreme weather–based investors, our rule will be to stick with the Upstream companies that have the least amount of their production located in hurricane alley. After all, we may not be able to predict which of the biggest loser companies (which have lots of production assets in hurricane alley) will be the unlucky company to get hit by the hurricane,

TABLE 6.2 Percent of Production in Hurricane Alley

Company	% of Production in Hurricane Alley
Southwestern Energy *(biggest winner)*	0%
Range Resources	0%
Continental Resources	1%
Devon Energy	2%
Whiting Petroleum	3%
Anadarko Petroleum	7%
Chesapeake	7%
Pioneer Natural Resources	8%
Cabot Oil & Gas	10%
Plains Exploration	13%
Noble Energy	17%
Apache Corporation	19%
EOG Resources	23%
Newfield Exploration	26%
Petrohawk Energy	33%
Denbury Resources *(biggest loser)*	38%

Source: Estimates based on public filings.

but we certainly can steer clear of all of those companies during hurricane season.

Let's take this one step further. Not only will we steer clear of the highest-probability losers, but we will buy the biggest winners because not only do they have very few assets in hurricane alley, but they also will be the major beneficiary when the price of natural gas skyrockets after the biggest losers are hit.

Although it is based on estimates only, Table 6.2 lists most of the major Upstream companies as well as an approximation of the percent of their total production that is located in higher-probability hurricane zones. This is our action plan table, showing in rank format the biggest winners and the biggest losers in the event a hurricane strikes the Gulf Coast region of the United States. Table 6.2 guides us to avoid the biggest losers and to buy the biggest winner(s).The stock ticker symbols for the major Upstream companies are shown in Table 6.3.

One of the most attractive features of investing in the North American independent natural gas/oil producers is that we have the choice of entering the futures market in natural gas directly, or, if we prefer, we can invest in the stocks or corporate bonds for the companies in this section because the companies in this section are not only all publicly traded but they are also flush with a large selection of corporate bonds as well.

TABLE 6.3 Equity Tickers for Companies Near Hurricane Alley

Ticker	Company
CHK	Chesapeake
EOG	EOG Resources
NBL	Noble Energy
NFX	Newfield Exploration
SWN	Southwestern Energy
PXD	Pioneer Natural Resources
DNR	Denbury Resources
WLL	Whiting Petroleum
HK	Petrohawk Energy
CLR	Continental Resources
RRC	Range Resources
COG	Cabot Oil & Gas
APA	Apache Corporation
APC	Anadarko Petroleum
DVN	Devon Energy
PXP	Plains Exploration

Source: Public filings.

Turn to Chapter 8, "Real-Life Examples: Execution, Results, and Timing," for examples of how these companies performed during Hurricane Katrina and guidance on how to respond to hurricanes that will show up in the future.

OIL

Recall from our earlier discussion on natural gas that oil is a globally traded and priced commodity, whereas natural gas generally is not. Oil, therefore, deserves a completely separate discussion.

For an extreme weather–based investor, oil is very attractive because it meets both of our ideal investment criteria: (1) it falls into the Jackpot category per our discussion of oil in Chapter 1; and (2) it has the property of geographic concentration, thus lending itself to the potential for global climate shocks. Interestingly, in the case of oil, historically it has been politically driven supply shocks and not weather-based supply shocks that impact this market. As mentioned earlier, commodities have no biases. A supply shock is a supply shock. Any type of major supply shock results in upward price pressure on the commodity.

Table 6.4 shows the geographic concentration of oil production around the globe.

TABLE 6.4 World Crude Oil Production

World Crude Oil Production	Percent
Russia	12%
Saudi Arabia	12%
United States	9%
Iran	5%
China	5%
Canada	4%
Mexico	4%
United Arab Emirates	4%
Brazil	3%
Kuwait	3%
Other	39%
World Total	**100%**

Source: eia.gov, 2009.

As shown, Russia, Saudi Arabia, and the United States hold the highest level of global concentration and, as such, will be our primary focus for oil. As we formulate our extreme weather–based action plan for oil, it is helpful to see how the stock price of the various global oil producers correlates with the raw price of crude oil. The correlation results are shown in Table 6.5. The equity ticker symbols for these oil-producing companies are shown in Table 6.6.

Interestingly, as shown in Table 6.5, the best correlation was for the Hess Corporation, at only 65 percent. In general statistical lingo, the interpretation here is that the price of crude oil alone can explain 65 percent of the changes in the stock price of the Hess Corporation. Why wouldn't it be much higher, like 95 percent? The answer is that there are many other things going on at Hess besides just selling higher-priced oil. Remember, Hess owns many retail-level gas stations. Gas stations generally do not like rapid increases in their input cost base, nor do the upstream refineries that are buyers of crude oil. Nevertheless, the correlation is positive and fairly high, meaning that as the price of oil climbs, so, too, will the stock price for Hess, statistically speaking.

The fact that the correlation is meaningfully less than 100 percent is the reason why we put the biggest winner in our action plan table as "Oil Futures" as opposed to the stock price of one of these companies. A futures-based exchange-traded fund (ETF) in oil would also be a suitable substitute as the biggest winner.

In summary, Table 6.7 shows our oil-based action plan in ranked format in the event a global supply shock hits the key oil-producing regions outlined in the table.

TABLE 6.5 Five-Year Stock Price Correlation with Crude Oil Price

Company	Five-Year Correlation of Stock Price versus Oil Price
Hess	65%
ConocoPhillips	61%
Statoil	57%
Petro China	54%
Rosneft	50%
Chevron Corporation	46%
Suncor Energy	46%
Continental	43%
Chesapeake	41%
Denbury	39%
Petrobras	37%
Eni Spa	36%
Royal Dutch Shell	34%
ExxonMobil	31%
Lukoil	28%
Whiting	28%
BP	25%
Repsol YPF	23%
Marathon Oil	23%
China Petroleum	10%

Source: Public pricing data.

Similar to our discussion on the independent natural gas producers, the global oil producers are publicly traded, which provides us with investment opportunities in the stock market. In addition, most of them are also flush with corporate bonds, thus providing us with additional flexibility in our financial market of choice. See Chapters 11 and 13, which cover the basics of investing in bonds and futures, for additional guidance in investing in these markets.

REMAINING STOCKS TO BENEFIT IN WIDESPREAD HURRICANES

During any severe hurricane there will be widespread damage to houses, buildings, and infrastructure. Although this guarantees an uptick in

TABLE 6.6 Equity Tickers for Oil-Producing Companies

Company	Ticker
Hess	HES
ConocoPhillips	COP
Statoil	STL NO
Petro China	857 HK
Rosneft	ROSN RM
Chevron Corporation	CVX
Suncor Energy	SU CN
Petrobras	PETR4 BZ
Eni Spa	ENI IM
Royal Dutch Shell	RDSA LN
ExxonMobil	XOM
Continental	CLR
Whiting	WLL
Denbury	DNR
Chesapeake	CHK
Lukoil	LKOH RM
BP	BP/ LN
Repsol YPF	REP SM
Marathon Oil	MRO
China Petroleum	386 HK

Source: Public filings.

TABLE 6.7 Crude Oil Action Plan

	Where's the Supply Shock? Russia	Where's the Supply Shock? Saudi Arabia	Where's the Supply Shock? United States
Biggest Winners (Ranked)	Oil Futures Non-Russians Hess Conoco	Oil Futures Hess Conoco Statoil	Oil Futures Non-U.S. Players Statoil Petro China
Biggest Losers (Ranked)	Rosneft Lukoil	Saudi Producers	U.S. Producers

Source: Public pricing data.

demand for such things as steel, chemicals, concrete, piping, copper wiring, and building materials, in general these effects take a much longer time to pan out. As a result, there are relatively few stocks that will have a direct and immediate positive impact. Therefore, we will leave this section out of the analysis as it relates to extreme weather–based investing.

CHAPTER 7

Global Climate Shock Number Five: Drought-Induced Timberland Fires

Drier-than-normal conditions often are blamed for the start of very large acreage timberland fires. When timberlands get destroyed, this represents a supply shock to the available logs for sale into the market. Now the log market is quite regional in nature, but in the event of a complete shut-off from a particular region of the country, other regions will step up to the table and fill the void, despite being in a more remote region. Logs from the West Coast of the United States, after all, are currently being shipped all the way to Japan and China, so they can travel long distances if the need exists. Many companies in the downstream forest products industry have sold off much of their timberland assets in the past decade, but there are four major public players who still retain significant timberland ownership that are worth watching: Weyerhaeuser, Plum Creek, Potlatch, and Rayonier. The exposure that each company has to log sales as well as the geographic exposure of their timberland assets is shown in Table 7.1.

When looking at Table 7.1 and deciding which companies would be the biggest winners and losers in the event of massive timberland fires in the northern, southern, or western United States, Table 7.2 is the result.

In the case of timberland ownership, even though a company has acreage in the north, they are still quite diversified from state to state within the north, so the timberland fire stock play is a bit trickier. The state to state diversity in timberland ownership dilutes the negative impact of a fire on a companies timberlands. As a result, the way to play the stock market on fires is to (1) keep this list handy so you are aware of who owns what, and (2) do not buy the stock of the company that falls into the biggest loser

TABLE 7.1 Timberland Companies

	Weyerhaeuser	Plum Creek	Potlatch	Rayonier
Logs (% of total revenue)	13%	49%	45%	13%
Geographic Split:				
Northern United States	0%	49%	17%	16%
Southern United States	63%	51%	27%	70%
Western United States	37%	0%	56%	14%
Total United States	**100%**	**100%**	**100%**	**100%**

Source: Public filings.

TABLE 7.2 Timberland Action Plan

	Where's the Fire? Northern United States	Where's the Fire? Southern United States	Where's the Fire? Western United States
Biggest Winners (Ranked)	Weyerhaeuser (13%)	All suffer	Plum Creek (49%)
Biggest Losers (Ranked)	Plum Creek (49%) Potlatch (45%) Rayonier (13%)	Plum Creek (49%) Potlatch (45%) Rayonier (13%) Weyerhaeuser (13%)	Potlatch (45%) Weyerhaeuser (13%) Rayonier (13%)

Source: Public filings.

category. In this case, our strategy is to avoid problem stocks. In the biggest winner category, although it is a general positive for the company, we will not buy the stock because of the extreme state-to-state diversity in timberland assets.

CHAPTER 8

Real-Life Examples: Execution, Results, and Timing

Now that we have reviewed all of the global climate shocks and extreme weather events, we will now take the next critical step and look at some specific, actual extreme weather occurrences and point out specific investment execution and timing strategies as well as results that would have been attained. We will do this to give you some real-world practice as a global climate shock and extreme weather–based investor. We will then end the chapter with a list of "The Rules of Extreme Weather-Based Investing."

HURRICANE KATRINA EXAMPLE

The first real-life example we will review falls under the heading of hurricanes. Specifically, we will look at the impact Hurricane Katrina had on the independent natural gas producers. The ranked action plan table was reproduced for convenience and is shown here as Table 8.1. As a reminder, this table lists in rank order the biggest winners and the biggest losers among the independent natural gas producers within the United States. The biggest winners are the ones with the least exposure to hurricane alley. The biggest losers are the ones with the greatest exposure to hurricane alley.

As shown, the two-week and the five-month stock returns were included in the table for every company in the list. To be more specific, the two-week return in the table is the return from four days prior to Hurricane Katrina to two weeks after Katrina. The pre-four-day figure

TABLE 8.1 Natural Gas Action Plan Table

Company	% in Hurricane Alley	Two-Week Return	Five-Month Return
NATURAL GAS PRICE	**NA**	**−11%**	**99%**
Southwestern Energy (Biggest Winner)	**0%**	**5%**	**106%**
Range Resources	0%	−14%	43%
Devon Energy	2%	−11%	37%
Whiting Petroleum	3%	−10%	36%
Anadarko Petroleum	7%	−8%	24%
Chesapeake	7%	−12%	69%
Pioneer Natural Resources	8%	−16%	21%
Cabot Oil & Gas	10%	−16%	38%
Plains Exploration	13%	−13%	22%
Noble Energy	17%	−3%	36%
Apache Corporation	19%	−12%	23%
EOG Resources	23%	−7%	50%
Newfield Exploration	26%	−10%	31%
Petrohawk Energy	33%	−20%	37%
Denbury Resources (Biggest Loser)	**38%**	**−15%**	**38%**

Source: Hurricane exposure estimates based on public filings.

refers to the typical amount of warning we are given by forecasters with hurricanes.

The results and conclusions are very interesting:

- Amazingly, even five months after Hurricane Katrina ended, the supply-side disruptions were still occurring. As a result, the price of natural gas over this five-month time frame was up by 99 percent. This translates into a direct win in the futures market. However, notice that in the first two weeks not only did most stocks decline (with the exception of the biggest winner) but the price of natural gas also declined. This is very exciting news. The laws of supply and demand dictate that when the supply side is disrupted, upward price pressure is the result. Instead, the price of natural gas and stock prices declined for the first two weeks in this case. This represents an excellent opportunity to improve your buy-in price. This is particularly true for a company like Range Resources, which has very little exposure to hurricane alley, and yet its stock price declined for the first two weeks. The message here for the next big hurricane is to watch carefully. If stocks and the futures market break lower, this simply means a better buy-in point, so be patient on the buy side.

- Notice also that the futures market would have beaten every single stock in the list over the five-month time period with the exception of the biggest winner, Southwestern Energy.
- When do you sell your hurricane-based investment? In general, when a supply shock correction takes place, and supply once again becomes available, market prices often go down again. Pay attention to the status of the supply, and sell before the supply availability gets completely repaired. But what if you decide to buy a biggest winner stock during the forecast period of the hurricane and then suddenly the hurricane decides to move out to sea instead of to land. If that is the case, then sell the stock immediately. As an extreme weather–based investor, our actions revolve around extreme weather events. If it turns out that the expected event does not occur, then get out of the investment.
- A key point here is that the upward price movement in natural gas was so strong that at the end of the five-month period, every single stock had a very strong positive return. That's fine because in the case of hurricane-based investing, our goal is to buy the biggest winners and to avoid buying the losers. The fact that even the biggest loser stocks made money is really of no consequence to us because our focus was on buying the stocks with essentially zero hurricane alley exposure.
- One of the most impressive conclusions of all from the Katrina example is the massive amount of time available for investors to get involved in a global climate shock or extreme weather–based investment. In this particular example, the price of natural gas continued to rise for five months, climbing 99 percent in value over this time frame. How is that possible? It is possible because the supply shock to the availability of natural gas was impacted for at least five months. Could anyone have guessed that supply would be impacted for at least five months? No, but that is precisely why the investment opportunity window is so wide—because of the massive unknowns associated with extreme weather events. As you will see, one of the most attractive investing features of extreme weather events is that their impact tends to last a long time. The longer the time the event impacts the market, the more time we have for an investment. Compare this vital and attractive feature of extreme weather–based investing with the more traditional method of investing, which looks strictly at earnings results. Once the company reports its earnings, the market gets the general picture for the company and there are very few unknowns left over. The stock price reacts almost immediately to the news of the earnings, whether or not they beat earnings estimates, and the investing opportunity is essentially gone. In the case of extreme weather–based investing, not only is time on our side because of the duration of the event and the fear associated with the lack of supply, but now we are in a far, far

better position than the average investor because now we have at our fingertips numerous action plan tables that we created throughout this book, which allow us to turn on an investing dime in reaction to any global climate shock in the world.

FLOODS IN EASTERN AUSTRALIA EXAMPLE

Recall from the action plan tables that in the event of extreme flooding in eastern Australia, the "met" coal market would see a massive supply-side disruption. According to our action plan tables, the biggest winners would be Walter Energy and Teck Resources, while the biggest loser would be MacArthur Coal. As it turns out, this is not a hypothetical event.

Very heavy rains began pounding eastern Australia in the second half of 2010. According to weather reports, in the three months ending October 31, 2010, rainfall in this region was up 400 percent more than the historic average. Figure 8.1 shows the stock price of Teck Resources

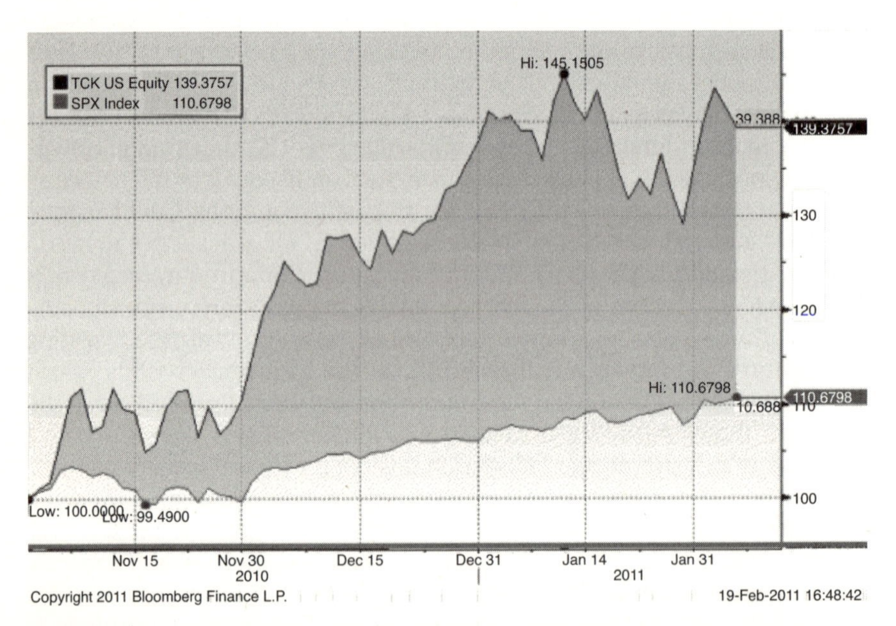

FIGURE 8.1 Teck Resources Stock Price versus S&P 500, November 2010 to February 2011
Source: Used with permission of Bloomberg Newswire Permissions.

versus the Standard & Poor's (S&P) 500 beginning on November 1, 2010. As shown, the stock price for the S&P 500 grew 11 percent over this time frame, but grew a whopping 39 percent for Teck Resources over this same time frame. Meanwhile, the biggest loser in this example, MacArthur Coal, showed a stock return of zero percent over this same time frame. Remember, MacArthur was not only located strictly in the eastern Australia flood region but also has 94 percent of its total revenue in this product category, so it is not surprising that Teck outperformed MacArthur by 39 percentage points over this time frame.

The basic conclusions from this example are as follows:

- The flooding in eastern Australia was a classic example of extreme weather–based investing. As shown in the stock price graph, as expected, Teck Resources, a biggest winner in our action plan tables, turned out to be a major winner in this event (we used Teck as opposed to Walter in this example because Walter in the relevant time frame was involved with the acquisition of another coal firm but for information purposes the Walter stock price was also up approximately 42% over this same time frame). Similarly, MacArthur Coal, as expected, turned out to be the biggest loser. Think about it—these results are not surprising. Of course, MacArthur Coal will be the biggest loser because nearly all of its revenue is in the met coal product line and its mines are also in this exact region of the world where the floods took place. Similarly, Teck would obviously win because it not only has the opportunity to gain market share but also to see its met coal price rise.
- In the case of met coal product pricing, there are no opportunities within the futures market because met coal does not trade in this market, so our investment opportunity rests in the stock market and the corporate bond market (see Chapter 11, "Opportunities in the Bond Market," for more detail on bond opportunities). However, the fact that there are no opportunities within the futures market does not change how this commodity reacted to a supply shock. The price of met coal has been on the rise relentlessly since the floods began in late 2010 straight through the writing of this book in the summer of 2011. Once again, this demonstrates the massive window of opportunity for investment in commodities that are hit by extreme weather events. The general rule is that the longer the number of days where supply is impacted, the larger the window of opportunity for investment.
- When is the best time to buy into the mine-flooding situation? In this particular example, the duration and the impact of the flood were so severe that you could have entered this opportunity even after the

floods became apparent at the early stages. The met coal price, as well as stock prices, continued to rise for many months even after the flood began. Again, the basic rule is the longer the duration of the supply-side shock, the more time we have to enter the market. The beauty of it is that although we cannot predict the duration of a supply side impact, we have seen time and time again that the impact is generally not short lived, which increases our window of opportunity. Remember, even if we enter an investment in the early stages of a supply shock before market prices have responded strongly and it turns out that the supply shock was just a minor event, simply sell the position immediately.

RUSSIAN DROUGHT EXAMPLE

In 2010, Russia saw the worst drought in more than 50 years. What began in the fall of 2009 with very low rainfall continued straight through 2010, with more and more regions of Russia receiving the "drought emergency" declaration. By August 2010, 23 regions of Russia were under this drought emergency (with the Volga District representing more than a quarter of total wheat harvest). See Figure 8.2 showing the price per bushel of wheat.

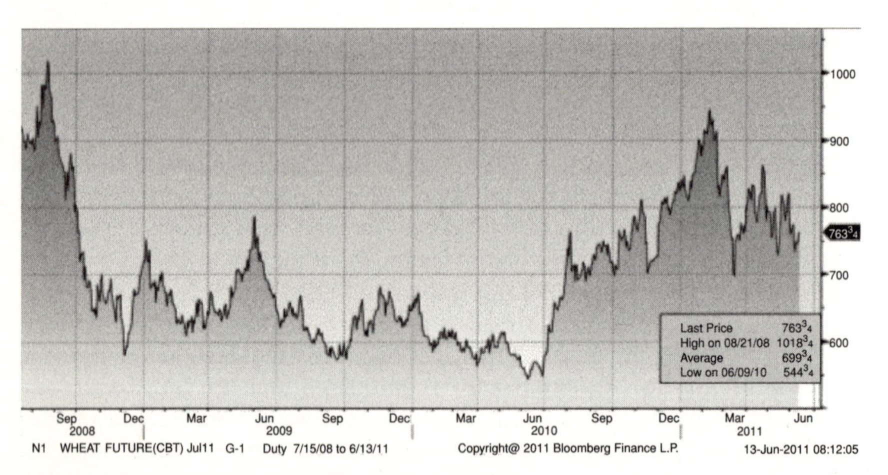

FIGURE 8.2 Wheat, July 2008 to June 2011.
Source: Used with permission of Bloomberg Newswire Permissions.

The drought situation in Russia is a very interesting example because it highlights the following key points:

- As discussed, drought conditions can persist for a very long time. In this case, drier conditions persisted for more than a year in Russia as more and more regions of Russia were declared regions under drought emergency. Remember from what we learned earlier, time is our friend as extreme weather–based investors. The longer the duration of the extreme weather event, the more time we have to allow our investment to appreciate.
- Beginning in the fall of 2009, low rainfall was the start of drought conditions. This condition persisted into 2010 through the spring and into the summer. The price pattern is very interesting. Notice that, despite the growing drought and the expectation of lower harvest levels, the price of wheat remained relatively stable into June 2010. But then in mid-summer, the Russian government announced its intention to institute a ban on wheat exports in August 2010. This is a fairly common practice these days as grain availability of supply gets tighter and tighter. Essentially, what happens is that the drought itself is a leading indicator of government action. This government action, in the form of a ban on wheat exports, acts as a double positive whammy for wheat prices because not only is the drought causing a global shortage of wheat, but now the Russian government exacerbates the situation by banning exports. In May 2010, it was announced that grain production was down by more than a third and yet pricing had not really responded yet in the wheat market. Had we invested in wheat futures in the May 2010 time frame, we would have earned a 37 percent return using the average value over the ensuing months, and, in fact, had we sold at the peak in February 2011, we would have earned a 63 percent return.
- How long do you hold the futures contract if you buy in before the price spikes? Of course, hindsight is 20/20, but ideally you would hold the contract until the cumulative drought conditions and/or the government intervention causes wheat prices to climb at which point you would continue to hold the position. This is a very personal decision because it is hardly crystal clear when or even if these events occur. You do, however, know that the longer the drought persists, the more likely you will see not only rising prices but also the increased potential for government intervention, which also acts as an additional supply shock to the world. While you are holding the position, you will need to continue to do your homework. You will need to keep a tab on drought conditions, public government opinions on the situation

and, importantly, on prices in the futures market. If, while waiting, you see prices declining, you may choose to exit the position, depending on your level of conviction on the severity of the drought and/or the expected actions of the government.

BLIZZARDS IN NORTHEASTERN UNITED STATES EXAMPLE

Recall from our discussion covering excess snow and ice that Compass Minerals (ticker CMP) was the stock of choice. It was the stock of choice because it not only had a heavy focus within the United States but also had a very high percentage of road salt in its revenue mix.

The winter beginning in late 2010 in the northeast region of the United States was a very severe one. There were multiple 12-inch-plus blizzards in the region. The demand for road salt is directly proportional to the quantity and frequency of ice and snow on the road surfaces.

Figure 8.3 shows the stock price of Compass Minerals as compared to the S&P 500 running from September 30, 2010, through February 28, 2011.

This graph provides a snapshot of the 2010/2011 winter. Over this time frame the stock market as a whole performed very well, returning 16 percent. However, Compass Minerals did even better, returning 22 percent

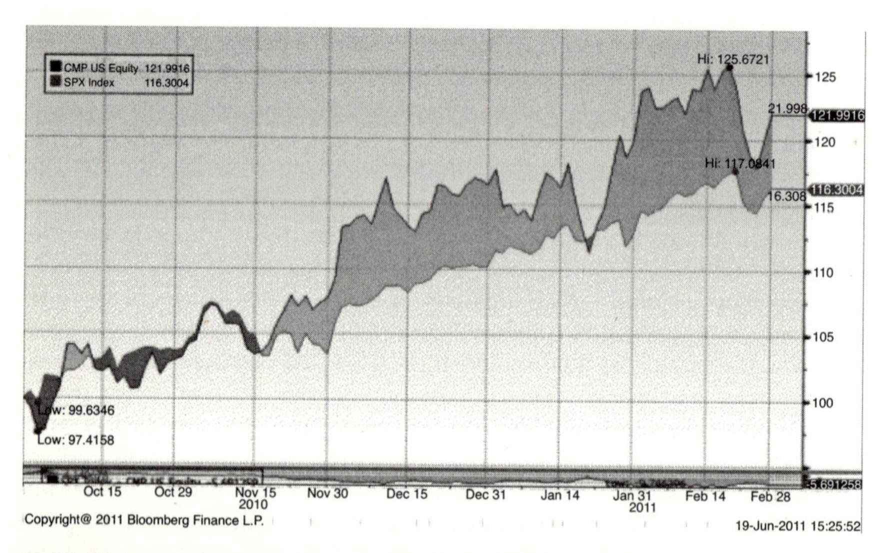

FIGURE 8.3 Compass Minerals (CMP) versus S&P 500 Index
Source: Used with permission of Bloomberg Newswire Permissions.

over the same time frame. The general conclusions from this extreme weather example are as follows:

- This example is different than many of the other climate situations covered in this book because the driver behind investing in Compass Minerals in this situation is not based on a supply shock. Rather, it is based on a large pickup in demand for road salt.
- The excessive snow and ice allowed a bet on the Compass Minerals stock, which ultimately outperformed the index by 600 basis points (i.e., 6 percent). Notice that the stock price of Compass and the S&P 500 Index tracked each other very closely over the September and October time frame, but when we hit the mid-November time frame, the Compass Mineral stock began to take off and outperform the Index.
- Interestingly, when looking at the Compass Minerals stock over the past eight seasons, you find that the stock ramps up heading into the winter and begins to decline into the warmer months year after year. An increase in the severity of the weather in the winter months will tend to heighten the uptick as we ramp into the winter.
- The method of playing the Compass Minerals stock is to get into the name before the November time frame, before the blizzards begin hitting the northeast.
- Similar to other extreme weather–based strategies we've talked about, if the weather does not produce the expected results and the market is not rewarding the effort, then sell out of the position immediately.

COCOA BEAN SUPPLY SHOCK IN EARLY 2011

I love this example because it points out that even supply shocks that are not related to extreme weather events still end up with the same basic result. The result is that the price of the commodity in question goes higher because the commodity is in short supply. The commodity in question in this example is the cocoa bean.

Recall from earlier chapters that cocoa beans have one of the most interesting stories from the standpoint of an extreme weather–based investor. It is interesting because approximately 37 percent of global cocoa bean production takes place in Cote d'Ivoire, a region on the northwest coast of Africa. Although a weather-based supply shock in this region would undoubtedly place upward pressure on the price of the cocoa bean, so, too, would a politically driven supply shock.

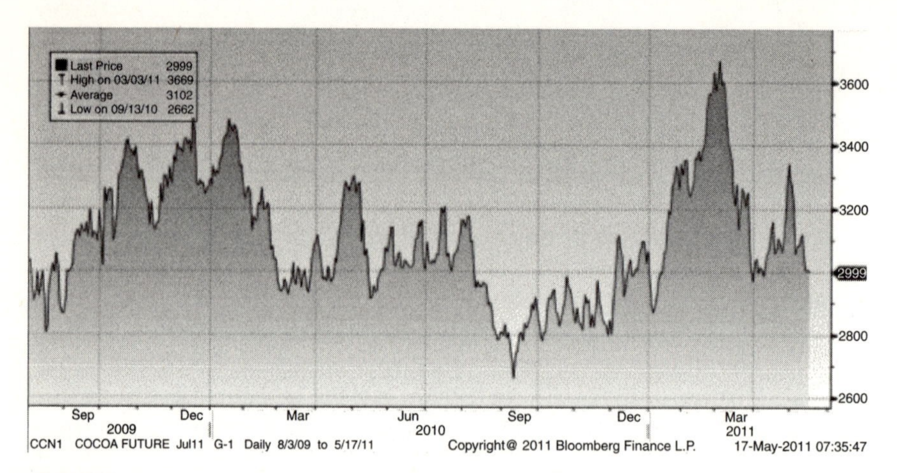

FIGURE 8.4 Cocoa, August 2009 to May 2011
Source: Used with permission of Bloomberg Newswire Permissions.

When the government of Cote d'Ivoire decided to put a ban on exports of cocoa beans out of the country for politically driven reasons (which they did in January 2011), it had the basic effect on price as a weather-based supply shock (i.e., prices go up).

The price of cocoa is shown in Figure 8.4. Notice the spike in the price of cocoa after the government temporarily banned exports in mid-January 2011, and notice in particular that even though the supply shock occurred in mid-January 2011, the price of cocoa kept rising for nearly two straight months after the ban was put in place, thus allowing investors time to capitalize on these events. If the futures contract were purchased at the time of the ban, an 18 percent return would have been realized over the next 7 or 8 weeks.

The good news in this example is that an 18 percent return was available. The challenging part is deciding when to get out. In this case, the issue was political in nature. Therefore, if you decide to invest based on politics, it behooves you to pay close attention to the politics to ascertain when to exit the position. In this particular example, it did not help global fears of weakening demand when the great Japanese earthquake struck on March 11, 2011. Granted, it is essentially impossible to predict earthquakes or other events that may negatively impact your investment, but in this example three general rules would apply:

- Whether a supply shock is from an extreme weather event or from a politically motivated event, the same rule applies on exiting. If at all possible, get out before the problem is fixed. If, for example, estimates are that a severe flood will take four to six months to fix, then get out considerably before the lower end of the range. Similarly, in the case of a politically driven supply shock, get out before the problem is fixed, if possible, or in this case before the ban is lifted.

- Now, let's assume that you are blindsided in your investment with an earthquake. If this occurs, and the price changes direction, then close out the position immediately.
- Similarly, let's assume that you do not get out in time before the supply shock is corrected (i.e., either the export ban is lifted or a hypothetical flood problem is corrected earlier than expected), then that officially ends the extreme weather investment. Your job then becomes to close out the position immediately.
- The beauty of it, as we have talked about time and time again, is that supply shocks, particularly the kind associated with extreme weather events, tend to be sticky, meaning they can last a long time. The longer, the better because this provides more opportunity for price appreciation in our investment whether that occurs in the futures market, the stock market, or the bond market.

A WORD ON CORN

I like to include corn in the discussion of relevant real-life examples because corn is the epitome of a Jackpot commodity at the present time. See Chapter 1 for a reminder as to why corn is such a solid Jackpot commodity.

Corn has such a strong fundamental story that the price of corn has been rising powerfully, despite the fact that there have been only minor weather events impacting the supply side. When a commodity falls into the Jackpot category, the supply/demand balance is already extremely tight, even without a global climate shock. In fact, the story is so compelling in many of these Jackpot commodities that they warrant investment even in the absence of an extreme weather–based supply shock. The key question is when to invest.

The next section of this chapter is a summary list of rules to follow as an extreme weather–based investor. These rules were derived from the results and conclusions obtained in the many examples provided throughout the book. Additionally, in light of the strength of the corn commodity, even absent an extreme weather–based supply shock, we will also talk about specific rules associated with Jackpot commodities in general.

THE RULES OF EXTREME WEATHER-BASED INVESTING

1. **Shock value.** To understand this point more clearly, we will create the definition of shock value.

Definition: Shock Value

The amount of time a supply shock persists. The higher the number of days, the greater is our window of time for entry into an investment. The higher the number of days, the greater our potential holding period of the investment.

As we have seen numerous times throughout this book, the shock value of extreme weather events can be quite long. Table 8.2 presents a few recent extreme weather events, as well as some other general extreme events in combination with a typical shock value.

As shown in Table 8.2, the shock value of extreme weather events can be quite long. This is the most attractive feature about extreme weather investing. The shock value is precisely why we have the opportunity to enter the market at the early stages of the supply shock and still have time to make money.

Interestingly, there is an enormous amount of time spent focusing on the quarterly earnings of companies by investors around the world. But think about it—what is the equivalent shock value of an earnings announcement? By the definition above, the shock value is less than one day for an earnings announcement, even if you include the amount of time it takes to read the quarterly news report and listen to the quarterly conference call by management. This is not a lot of time. In other words, attempting to benefit from quick stock price appreciation after the quarterly earnings report is released is too late to the game.

TABLE 8.2 Shock Value Examples

Extreme Event	Shock Value
Hurricane Katrina	6 months
Russian drought	1 year
Eastern Australian floods	3-6 months
Blizzards in series	3 months
Mining strike	30 days
Earthquake	1-6 months

Source: Estimates based on historical events.

2. **Commodities are not biased.** Although the primary focus of this book is on extreme weather–based investing, the fact of the matter is that commodities have no bias. A supply shock is a supply shock. It does not really matter what causes the supply shock, whether it is from an extreme climate event, an earthquake, a large strike, or a politically motivated event, the end result is fundamentally the same. The available supply of the commodity gets cut off. The thing that matters is the shock value, the number of days the supply shock persists as well as the level of geographic concentration of the commodity in question. See the next point for more on geographic concentration.
3. **Have the biggest winner and biggest loser tables at your fingertips.** Throughout this book, we have generated and discussed numerous tables showing the biggest winners and the biggest losers for dozens of commodities mapped against essentially every conceivable extreme weather event. The smartest among the readers will have trouble remembering all of the tables, and therefore it makes sense to have the tables sitting at your fingertips, available for when the next extreme weather event strikes. The tables will not only provide the specific names of the companies and financial markets that will win and lose, but will also detail the global geographic concentration of each commodity. This is very important. For example, a supply shock to the cocoa bean in the Cote d'Ivoire region is one of the most attractive combinations because this particular region produces a whopping 37 percent of the global supply of cocoa beans. If this available supply were cut off, particularly for a long period of time, it would move markets meaningfully.
4. **Jackpot commodities usually beat Big Problem commodities.** All other things being equal, Jackpot commodities usually have a stronger reaction to a supply shock. This is true because the supply/demand fundamentals in a Jackpot commodity are already very tight. The supply shock exacerbates an already-tight condition. An exception to this general rule is the case where a Big Problem commodity sees an excessive shock value (number of days the supply shock persists). For example, during Hurricane Katrina, the market price for natural gas rose by 99 percent over the next five months post Katrina. A longer shock value continues to put upward pressure on commodity prices. It must be noted, however, that in the spring of 2005 when Katrina occurred, the commodity, natural gas, was more in the Neutral category as opposed to the current Big Problem category primarily because the great shale gas discoveries were not yet in full swing. Nevertheless, the original premise that Jackpot commodities generally win still holds true.

5. **Deciding when to enter an investment.** First of all, be patient. As an extreme weather investor, the question is not *if* more extreme weather events will hit but *when* they will hit. We want to wait for the "perfect storm," so to speak. Specifically, the investments of greatest interest are the ones that meet most if not all of these criteria:
   - **a.** *Geographic concentration.* The higher the geographic concentration, the better. For example, cocoa beans beat nickel metal in terms of global geographic concentration of supply, as we have already covered. Similarly, corn beats zinc. As you can see, having the biggest winner and biggest loser tables at your fingertips is critical to assess the geographic region where the global climate shock occurs.
   - **b.** *Shock value.* Look at historic shock values (typical number of days a supply shock persists) for the type of extreme weather event that is occurring at the time of your consideration. Remember, time is our friend. The higher the shock value, the better. Also remember, as we talked about earlier, sometimes a shock value has a beneficial, second-layer effect that causes the shock value to become even more extended. For example, it is becoming fairly common practice that when a particular grain sees a supply shock and prices for that grain are climbing, the government of the country that was hit institutes an export ban on the commodity. Their intention is usually pure in that they want to make sure their own people have access to the material. However, the resultant effect is a second-level supply shock. In many cases, the export ban is more devastating to supply than the original drought. As a result, the drought itself can be a leading indicator to government action and an additional price spike in the relevant commodity.
   - **c.** *Timing.* As is always the case, timing is important and challenging but gets easier with higher shock values. A key point to remember is that according to basic economic theory, a supply shock will typically lead to higher prices. Interestingly, on occasion, the price of a commodity may drop immediately following a supply shock (as was the case with natural gas over the first two weeks post Katrina, but later increased in price by 99 percent over the subsequent five months). Often, this represents a better buy-in price. So, immediately following the initial stages of the supply shock, do not just dive into the investment. Watch the direction pricing takes. If it declines initially, there may be a better buying opportunity once the price settles before it again begins to move higher.
   - **d.** *Optimum market cycles.* Commodity prices, as a rule, are quite volatile. Even under continuing good supply/demand fundamentals,

the price of commodities can still swing. Ideally, we like to enter a commodity position after a recent pullback in pricing, particularly if the commodity we are dealing with is a "jackpot" commodity and will likely bounce back at some point. Entering an extreme weather investment that also coincides with a recent pullback in commodity prices is ideal. Similarly, entering the stock market during times when the market price-to-earnings multiples are at recent lows also helps support our positions.

6. **When to exit an investment.** The following are among the numerous reasons to exit an extreme weather–based investment:
   a. *Ideally, close out the position before the supply shock ends.* It is often difficult to guess when a supply shock will come to an end, but there are clues. For example, in an extreme flood situation (as was the case with metallurgical coal in eastern Australia), market participants will state publicly how long of an interruption they expect in the supply availability. You must pay attention to this type of information so that you can form your own opinion after listening to others. If most people think that the flood supply shock will persist for four months, then plan on exiting the position before that occurs. The number one place to focus your attention is on the company that is the biggest loser! They are the ones that are living in the devastated region and therefore they are in the best position to assess when the problem will be fixed. They also are likely to be the most vocal market participant precisely because they are trying to allay the fears of their own investors.
   b. *Exit immediately forecast plays that do not pan out.* For example, if you enter an investment based on a severe weather forecast because it meets all of the "perfect storm" criteria discussed earlier, but then at the end of the day the forecast does not pan out, then sell the position immediately. These plays are event driven in nature. If the event does not occur as planned, then exit the position immediately.
   c. *Thoughts on selling too early.* Let's assume that we have a Jackpot commodity situation that is seeing an extreme weather event that most people think will have an estimated shock value of nine months. You enter the position and happily watch not only the commodity price go higher but also your chosen financial instrument go higher (i.e., stock/bond/futures contract). So let's say you are at month four of the investment and you have already made a respectable return. There is nothing wrong with selling the position at this point, even though it looks like pricing may rise for quite a while going forward. Greed is always a dangerous game. However,

some of you may choose to hold the investment for a little longer. If you decide to hold out a bit longer in this situation, remember, three things should remain intact: (1) there should still be plenty of time remaining before the expected end of the supply shock; (2) the extreme weather event has occurred exactly as expected; and (3) the technicals of the commodity pricing are holding up well. In other words, the first two criteria may be intact, but if commodity prices appear to have given up, then it is time to get out.

7. **Seasonal plays can work on both sides of the coin.** This particular point works best when dealing with Jackpot commodities when *buying* into a seasonal climate shock. Similarly, it works best when dealing in Big Problem commodities when *avoiding* certain biggest loser companies during seasonal bets identified in our action plan tables. An example of both of these scenarios will help clarify.
    a. ***Buying** a security before a particular season starts.* As discussed earlier, the commodity salt falls into the jackpot category. In addition, the maker of road salt, Compass Minerals, often sees its stock price ramp up as it enters the winter months and come back down after the winter ends. It has repeated this pattern in the past. Because salt is a Jackpot commodity at present and is a seasonal play, and because extreme winter weather has been the norm most recently, buying into this name prewinter can be a successful plan, as it has in the past. As stated before, if, as the winter progresses, the expected blizzards do not arrive, then this is grounds to exit the investment.
    b. ***Avoiding** particular companies before a season starts.* Let's use the independent natural gas producers during hurricane season as our example. As we learned in our action plan tables, certain companies have a greater percentage of their assets located in hurricane alley. The point in this example is simply to avoid buying these loser companies during hurricane season. Instead, the focus would be on buying the biggest winners during hurricane season that have very little asset exposure to hurricane alley.
8. **Avoid the losers.** Here, the word *losers* refers to two different types of losers. The first type of loser is the company that is buying the commodity whose price is spiking. For example, when the price of sugar is spiking, avoid buying Hershey. An entire table of buyer-customer relationships is included in Chapter 9, "Playing Both Sides of the Coin." In each case, whichever commodity is spiking, avoid the customer buying that commodity. We avoid buying the customer because we know ahead of time that they will be experiencing some headwinds in terms of input cost pressure, and therefore their stock price may struggle.

The second type of loser refers to the action plan tables laid out throughout the book. This second type is the global *competitors* of the biggest winners. They are the ones that were physically hit by the global climate shock and as a result will not be able to ship product. They also will not be able to sell product at the new higher price level resulting from the supply shock.

Also see Chapter 9 for additional opportunities for longing (buying) and shorting (selling) the winners and the losers simultaneously.

9. **Sometimes it is OK to buy the loser.** Throughout this book, we have created and discussed the biggest loser and the biggest winner action plan tables. Is it ever acceptable to buy a loser in the table? This is a critical point and is among the finer details, but must be used with caution. There is one circumstance where buying the loser is acceptable. Let's assume we are dealing with a Jackpot commodity. Let's further assume that there is a global climate shock in a key geography of a producer of this commodity. This combination of events places this producer into the biggest losers list. Although it is true that this company will run into some hard times because they will not be able to ship their product and they also will not benefit from the commodity price spike, longer term this could be an opportunity. Again, a key assumption here is that we are dealing with a Jackpot commodity and a strong company in general. Given the strong fundamentals, there is a reasonable chance that the stock price of this company may eventually bounce back when the supply shock ends, after initially falling during the extreme weather event. This is more of an advanced technique in extreme weather investing.
10. **A sun shower in Phoenix will not cut it.** This semicomical rule is stressing the point that what we are talking about here are *extreme* weather events. This is why we provided numerous examples to help give you the flavor of the magnitude of events that move the needle, thus causing financial markets to respond. You don't need to buy a futures contract every time you hear rain in the forecast. Remember, this is the primary reason we provided the numerous biggest winner and biggest loser investing tables. They provide you a quick snapshot of the key potential extreme weather investing opportunities that you should always keep at your fingertips.

    As an extreme weather–based investor, it is time for you to become much better at geography. Let's say we hear that massive frost conditions hit the south. Immediately and correctly, you check the chapter of the book covering frost and quickly determine that oranges hate frost and this could be a potential investment in orange futures. You further discover from our "biggest winner/loser" tables that the United

States is a key global producer of oranges and more specifically that Florida produces 80 percent of the oranges in the United States. So if "the south" means Texas or some other incorrect region, that does not help our orange play because Texas is not a major producer of oranges like Florida. Now, let's assume that massive frost conditions are hitting Florida. Now we are getting warm on a potential investment. Your job is to do some research and to determine exactly where the frost conditions are hitting and specifically whether they are hitting the orange-producing regions of Florida. If it turns out that this extreme weather event is hitting the key regions, then you may have a good opportunity on your hands. This same detailed geographic analysis must take place for each extreme weather–based investment you make.

11. **Sometimes corporate bonds are a longer-term play.** As discussed in Chapter 2, "Where to Invest: Stocks, Bonds, or Futures?," the percent return we earn in corporate bonds is very often lower than the returns we can achieve in the stock market or in the futures market. However, bond pricing is generally more stable and therefore quite attractive from a risk standpoint. Even in the case where we have an extreme weather event that drives up the stock price of a company significantly, the associated corporate bond price may not respond as sharply. Nevertheless, the extreme weather event may represent a good buy-in point for one of our biggest winner corporate bonds. In addition, if the biggest winner company also happens to be producing a commodity in the Jackpot category, our purchase of this corporate bond may turn out to be a relatively long-term investment even after the global climate shock or extreme weather event has gone away. For additional guidance on the key factors in selecting a corporate bond, see Chapter 11, "Opportunities in the Bond Market."
12. **Near-term carbon tax–based investments are limited.** When it comes to carbon tax–based investing, the near-term opportunities are limited. The impact of a carbon tax on the companies that produce carbon emissions will in many cases take years to occur. In addition, in many cases the carbon-emitting company will simply pass along the cost of this tax to their downstream customers and therefore may have relatively minor impact on its earnings. For these reasons, as extreme weather–based investors, we will keep the focus on the low-hanging fruit investment ideas discussed throughout the book. See more on carbon tax in Chapter 9.
13. **The preferred timberland fire "play" is to avoid the impacted companies.** Throughout the book, we have generated numerous biggest winner and biggest loser action tables. Generally speaking, our goal has been to buy the biggest winners. In the case of timberland

fires, however, our goal is to simply avoid the biggest losers and not to buy any of the timberland owners. The avoidance strategy is based on the extreme state-to-state diversity in timberland ownership. See Chapter 7, "Global Climate Shock Number Five: Drought-Induced Timberland Fires," for additional guidance in this area.

14. **Sometimes a demand spike and not a supply shock drives results.** Occasionally, an extreme weather–based event causes the demand side for a commodity to spike upward, as opposed to the dominant theme of the book with supply shocks as the key driver. As discussed earlier, road salt demand is a great example of this phenomenon, which occurs during heavy snow and ice events.
15. **Futures markets win because they don't care where the shock occurred.** Whether the global climate shock occurred within a key mining location globally or whether it was related to a particular global agricultural commodity, the futures market always wins, regardless of the geography of the supply shock. This makes an investment in futures markets more straightforward than stocks because futures markets care only about the price of the commodity in question. Stock prices by contrast take into account many additional factors such as cost inputs, managerial changes, and strikes, among many other factors.
16. **Multiple second-tier-type investment opportunities exist.** Throughout this book, numerous investment ideas have been presented. Not all of these ideas are the primary beneficiaries of a supply shock, but rather still exhibit a strong stock price correlation to the companies that are the primary beneficiaries to the supply shock. An example includes the equipment makers feeding both the mining and the agricultural commodity producers. These second-tier investments can help to diversity your portfolio, but our primary focus will remain on the direct beneficiaries of extreme weather events and global climate shocks.
17. **Jackpot commodities are often great investments even absent extreme weather events.** We are at a point in history where a wonderful collision is taking place in the commodity world. On one side, we have strong global demand, particularly in regions such as China, India, Brazil, and Russia. On the other side, we have scarcity of supply. Interestingly, this is the basic story for many commodities, thus making the sector quite attractive in terms of its core fundamentals. As a result, a word on commodities as a general investment even absent global climate events is called for.

    As mentioned previously, despite strong fundamentals, even Jackpot commodities will periodically see price swings. This is true whether we are talking about pricing in the relevant futures contract or about the stock price of a producer of the relevant commodity. The simplistic conclusions, therefore, are (a) that "jackpot"-type commodities in general should be part of a long-term investment strategy, and (b) given the price swings that occur, we should seek to enter our commodity-type investments at a point in time after the market prices pull back.

CHAPTER 9

Playing Both Sides of the Coin

In the world of stock picking, where there is a winner, there must be a loser; and if there is a loser, then there must also be a winner somewhere. We will use this fact to play both sides of the coin. What I mean by playing both sides of the coin is to buy the winner stocks and to sell the loser stocks simultaneously. How do we do that? Read on.

A couple of examples always help to clarify things. Rising sugar prices is great news for the sugar farmer (and the owners of sugar in the futures market) but bad news for the buyers of sugar like Hershey and Coca-Cola. So the winner in this case is sugar prices, and the loser in this case is the stock price for Hershey and Coca-Cola.

Another example, when a Big Problem commodity such as natural gas sees its price collapse as a result of the massive increase in supply due to the new horizontal drilling techniques, for example, that is bad news for the producers of natural gas but fantastic news for the buyers of natural gas including chemical companies and fertilizer makers.

This win/lose interplay does not just have to take place between the sellers and the buyers of materials. *The win/lose interplay also occurs between competitors, especially and in particular during supply shocks associated with global climate change.* Recall the numerous biggest winner and biggest loser lists that we have created throughout this book. Notice in those lists, the losers were not buying anything from the winners. They were competitors. Regardless of whether the win/lose interplay occurs between buyers and sellers or between competitors, we can make money either way by buying the winners and selling the losers *simultaneously.*

Before we get into detailed examples proving the power of this concept, let's first talk about why in the world we would buy and sell two different coal companies, for example, at the same time. First of all, yes, it is possible to sell a stock you don't even own. It's called *short selling.* It's simply the reverse of what investors normally do. Normally, investors buy a stock first, and then at some point in time they choose to sell the stock. We do this with the hope that the stock price will go up, so that we buy low and we sell at a higher stock price and therefore make some money. Short selling is simply the reverse. In this case, we believe that the stock price might go lower, so we sell the stock today and buy it a lower price later (behind the scenes your stock-trading account technically borrows the stock from someone else and loans it to you so you can sell it, and after the stock price moves lower, we go into the market and buy the stock at the lower price so that we can return the borrowed stock). At the end of the day we accomplished the same thing as we normally would do—that is, we buy low and sell high. The only difference is that we had to sell first here because we believe the stock is going lower. This trade is easily executed online with your trading account because all of the behind-the-scenes "borrowing" of stock takes place automatically and without your having to be involved. You simply need to click the "sell" and "buy" buttons in your trading account.

OK, back to the question: why buy and sell two coal companies, for example, at the same time. I mean, if the market as a whole goes higher, doesn't that imply that the stock you bought will earn money but the stock you sold short will lose money and you will net out to zero-dollar profit? That would be true if the coal companies we picked were each essentially the same type of coal company. However, that is not our goal. We want to buy the winners and sell the losers, which have been painstakingly laid out in this book. In this way, we may be able to make money on both sides of the coin; and, by the way, there is one additional very large benefit of buying and selling simultaneously in the market. Let's say the entire stock market chooses to take a violent crash downward on any given day. If we had bought only the winner stock and the market crashed, it may very well overwhelm the "winner" features of that stock and we could lose lots of money. However, if we buy and sell simultaneously in the market and the market crashes, yes, we would lose money on the stock we bought, but we would make money on the stock we sold short. In other words, we are protected, like an insurance policy, from severe declines in the market. So even under the dire scenario of the market's crashing, we would expect to still make a profit because the winner we bought would be more resilient to a crash, and the loser we sold would be particularly sensitive to a crash because it is also dealing with a very negative global climate change situation, such as its coal mines are flooded. OK, now that we are clear on the simultaneous buying and selling of stocks, let's look at some examples to make the point clear and to make money on the biggest winners and biggest losers tables in this book.

MET COAL EXAMPLE

The metallurgical ("met") coal example is well timed because at the time of writing this book, the eastern Australian coal mines are under water from floods. This is a perfect example of weather-induced supply shock to this market. So how do we use the winner/loser table that was pulled from our earlier discussion (referenced here as Table 9.1)?

The first thing we do is find the geography in Table 9.1 that flooded, which in this case is eastern Australia. So once that is identified we find the biggest winner and the biggest loser. Remember, both the biggest winners and the biggest losers are ranked. Therefore, you will get the biggest bang for your buck by buying the best of the winners (i.e., Walter Energy) and selling the worst of the losers (i.e., MacArthur Coal). As expected, over the November 2010 (just before the flood) to February 2011 time frame, Walter saw its stock rise by 42 percent, followed by Tech at 39 percent (second best winner), while MacArthur (the worst of the losers) showed

TABLE 9.1 Met Coal Action Plan Table

	Where's the Flood? Eastern Australia	Where's the Flood? Western Canada	Where's the Flood? Eastern United States
Biggest Winners (Ranked)	Walter Energy (80%) Tech Resources (50%) Alpha Natural (13%)	MacArthur (94%) Xstrata (15%) BHP (12%) Alpha (13%) Anglo (12%) Rio Tinto (10%) Peabody (10%)	MacArthur (94%) Tech Resources (50%) BHP (12%) Xstrata (15%) Anglo American (12%) Rio Tinto (10%) Peabody (10%)
Biggest Losers (Ranked)	MacArthur (94%) Xstrata (15%) BHP (12%) Anglo American (12%) Rio Tinto (10%)	Walter Energy (80%) Tech Resources (50%)	Walter Energy (80%) Alpha Natural (13%)

Source: Public filings.

a stock price change of zero percent over the same time frame. You might say to yourself, "So why bother selling short the MacArthur stock since it didn't decline?" The answer is that you have a choice. If you are feeling particularly bullish about the market, then simply buy the winner and call it a day. If you want the added insurance to protect you from an unexpected collapse in the stock market as a whole, then buy the winner and sell the loser and sleep more soundly, and in this case still make a 42 percent return on your investment! At the time of this writing Walter was buying another coal company so if we chose to go with Tech instead of Walter at this time, we would have made a net return of 39%.

IRON ORE EXAMPLE

The beauty of the met coal example is that the global climate change extreme weather shock has already taken place, so we had some concrete numbers to study. The beauty of this iron ore example that follows is that we have done the preparation work and are ready to pull the trigger when extreme flooding blocks off specific iron ore mines. Let's look at the winner/loser table for iron ore, shown here as Table 9.2.

As shown, when the day comes when the western Australian iron ore mines are under water from cyclone flooding, be ready to watch the stock price of Cliff's Natural Resources and Vale climb while watching the stock price for Fortescue (FMG) drop. This is a perfect long/short pairing (long means to buy the stock and short means to sell the stock—I know, more investor lingo). That is, in this case, buy Cliffs and sell FMG. Again, if you are feeling particularly bullish about the market as a whole, then you

TABLE 9.2 Iron Ore Action Plan Table

	Where's the Flood? Western Australia	**Where's the Flood? Brazil**	**Where's the Flood? India**
Biggest Winners (Ranked)	Cliffs Natural (89%) Vale (61%) Vedanta (15%)	FMG (100%) Cliffs Natural (89%) Rio Tinto (28%) BHP (21%) Vedanta (15%)	FMG (100%) Cliffs Natural (89%) Vale (61%) Rio Tinto (28%) BHP (21%) Vedanta (15%)
Biggest Losers (Ranked)	FMG (100%) Rio Tinto (28%) BHP (21%)	Vale (61%)	

Source: Public filings.

may choose to bypass the insurance of selling FMG and simply buy the winner(s).

COPPER EXAMPLE

Similar to the iron ore example, we would use the winner/loser table in the same manner as we did for both iron ore and met coal, shown here as Table 9.3. Simply wait for the extreme weather supply shock to hit Chile or Peru and take the directed action by buying the winner and selling the loser.

FARMLAND DROUGHTS/ FLOODS EXAMPLE

In the case where corn or soybean acreage gets impaired by severe geographic drought (like the recent drought in Russia with wheat), grain prices spike, as we talked about, and the key farm inputs also benefit, including fertilizers, seed makers, and agricultural chemicals. However, if you notice in the winners/losers grid in Table 9.4, there are no losers on which we can play the opposite side of the long/short trade. That is because the only loser in this scenario is the specific farmland that is flooded or is suffering from drought, and we have no market to play that. However, never fear—creativity always prevails. You have two ways to play this: (1) simply buy the best winners, as shown in the table; or (2) if you want the protection of having a short sale included in the trade, you can always short the Standard & Poor's (S&P) 500 (ticker SPY) as a whole, called *shorting the market.*

TABLE 9.3 Copper Action Plan Table

	Where's the Flood? Chile	Where's the Flood? Peru
Biggest Winners (Ranked)	Copper Futures Market Southern Copper (70%) Freeport McMoRan (80%)	Copper Futures Market Freeport McMoRan (80%)
Biggest Losers (Ranked)	Xstrata (40%) Anglo (39%) Rio Tinto (14%) BHP (13%)	Southern Copper (70%) Xstrata (40%) Anglo (39%) BHP (13%) Rio Tinto (14%)

Source: Public filings.

TABLE 9.4 Farmland Action Plan Table

	Farmland Droughts and Floods
Biggest Winners (Ranked)	Commodity Grain Futures Markets (100%)
	Monsanto (100%)
	Syngenta (100%)
	Potash Corp (100%)
	Mosaic (100%)
	Agrium (100%)
	CF Industries (100%)
	Yara (100%)
	Uralkali (100%)
	Silvinit (100%)
	Dupont (29%)
	FMC (40%)
	Chemtura (13%)
	Dow (10%)
	BASF (6%)
Biggest Losers	The farmlands seeing floods/droughts

Note: Percent of total revenue in the fertilizer OR seed OR Ag chemical markets.
Source: Public filings.

ADDITIONAL PAIR TRADES

As mentioned earlier, there are other pair trades available, but these trades carry more risk because they involve two completely different industries. We will walk through an example so you can get the flavor for this type of trade, and I will provide some caveats. But before venturing out into these trades, I would recommend studying Chapter 10, "Basic Principles of Commodity Investing," to beef up your understanding of stock valuation.

We will walk through the "sugar versus chocolate" example next. We will then provide numerous other pair trades in Table 9.5 that are potentially available, along with final recommendations on these types of pair trades in general.

Sugar versus Chocolate Example

As mentioned earlier, sugar is a key input into chocolate production. As shown in Figure 9.1, over the past six months the price of sugar is up 59 percent, but the stock price of Hershey is up only 12 percent compared

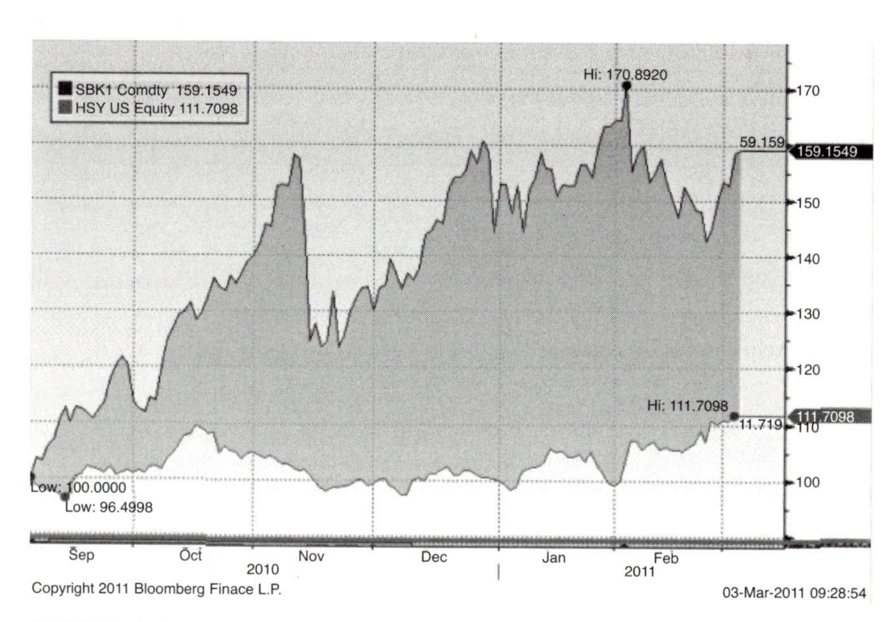

FIGURE 9.1 Sugar Price versus Hershey's Chocolate Stock Price, September 2010 to February 2011
Source: Used with permission of Bloomberg Newswire Permissions.

to the S&P 500 as a whole, which grew 20 percent over the same time frame. This indicates that sugar prices may have been weighing on the Hershey stock price (ticker HSY) over this time frame.

In order to give you a taste of the numerous opportunities in these types of pair trades, Table 9.5 is provided. The relevant stock tickers for these names are included in Table 9.6 for your convenience.

As shown, there are dozens of opportunities in Table 9.5. The question is how do we use the pair trades demonstrated in these tables? As extreme weather–based investors, our top priority is to invest in price spikes associated with global climate shocks. This was the purpose for generating the numerous action plans throughout the various chapters. The difficulty in these types of pair trades (i.e., where there is a buyer and a seller involved) is that the customer in many cases may actually be able to deal effectively with the spike in the cost of his key purchased commodity. He may be able to increase his own prices downstream to fully pass through his costs. In that case he would not be a loser at all. In some cases he would not be able to increase his price sufficiently and his profits and his stock price would consequently get squeezed. Whether he can do it or not depends on the health of the industry in which he operates. Given the challenge associated

TABLE 9.5 Addititional Pair Trade Opportunities

Commodity Spike	Category Negatively Impacted	Examples of Who Loses
Sugar	Soda and Chocolate	Hershey, Coca-Cola, Pepsi
Corn	Meat, Food, Fuel	General Mills, Kellogg's, Pilgrim's Pride, Coca-Cola, Pacific Ethanol, Pepsi, Tyson, Sanderson Farms, Hormel, Smithfield
Coffee beans	Coffee	Sara Lee, Starbucks
Cocoa	Chocolate	Hershey
Wheat	Food/Beverages	Anheuser Busch, Heineken, Kellogg, General Mills
Natural Gas	Chemicals/Fertilizers	CF Industries, Yara, Westlake
Nickel	Stainless Steel	AK Steel
Met Coal	Nonintegrated Carbon Steel	AK Steel
Iron Ore	Nonintegrated Carbon Steel	AK Steel
Soda Ash	Glass Makers	Owens Illinois, PPG
Lead	Car Battery Makers	Exide, Johnson Controls
Cotton	Apparel	Hanes, Nike, The Gap, American Eagle, Aeropostale
Steel	Car Makers/Construction	General Motors, Ford, Honda
Aluminum	Car Makers/Construction	General Motors, Ford, Honda
Oil	Airlines, Plastic converters	SW Airlines, Delta, Jet Blue, United Continental, Clorox
Beef	Retail beef sellers	McDonald's, Jack in the Box

Source: Public filings.

with being able to predict his success in raising his own prices to compensate for his increased costs, the best way to use the names in the table is by employing what I call *stock avoidance*.

For example, if we discover that the Cote d'Ivoire region of the world sees a major global climate shock, then we would instantly know based on our action plan tables that the price of cocoa will skyrocket because nearly 40 percent of the world's cocoa is made in Cote d'Ivoire. The investment we would make is in the cocoa futures market. Although it is true that the stock price of the Hershey Company may very well suffer from this event, it doesn't mean that we should necessarily short (i.e., sell) their stock. Instead, I am recommending that we use this pair trade table as a guide to stock avoidance. So, in this particular example, our action plan within the pair trade table is to avoid buying the Hershey Company stock because we know there is a good chance their stock is about to underperform

TABLE 9.6 Equity Tickers for Various Pair Trades

Company	Stock Ticker	Company	Stock Ticker
Hershey	HSY	PPG	PPG
Coca-Cola	CCE	Exide	XIDE
Pepsi	PEP	Johnson Controls	JCI
General Mills	GIS	Hanesbrand	HBI
Kellogg's	K	The Gap	GPS
Pilgrim's Pride	PPC	American Eagle	AEO
Smithfield	SFD	Aeropostale	ARO
Pacific Ethanol	PEIXD	General Motors	GM
Tyson	TSN	Ford	F
Sanderson Foods	SAFM	Honda	7267 JP
Hormel Foods	HRL	Southwest Airlines	LUV
Sara Lee	SLE	Jet Blue	JBLU
Starbucks	SBUX	United Continental	UAL
Anheuser Busch Inbev	ABI BB (BUD)	Delta	DAL
Heineken	HEID NA	Clorox	CLX
CF Industries	CF	McDonald's	MCD
Yara International	YAR EU	Jack in the Box	JACK
Westlake Chemical	WLK	Owens Illinois	OI
AK Steel	AKS		

Source: Public filings.

compared with the stock market as a whole. Think of this pair trade table as a guide to avoiding land mines. By having this table at our fingertips, it will help to avoid inadvertent mistakes such as buying a stock that will take a direct and negative hit from an extreme weather event.

Thoughts on Carbon Tax

The United States is late to the game in terms of adopting the Kyoto Protocol, which is aimed at reducing the level of carbon dioxide and other greenhouses gases in the atmosphere. Many companies, in anticipation of a government carbon tax, have already begun preparing by spending the money to lower their emissions, and many have not. In some cases it is very difficult, if not technically impossible, to eliminate all of their emissions. Power plants, for example, will likely see some form of carbon tax because within the United States the number one fuel source for electricity generation is thermal coal. Chemically, coal is predominantly carbon. When the power plant burns this fuel to help generate electricity, the emission product is carbon dioxide. These companies are among the highest emitters of carbon dioxide in the country. Yet when it comes to betting on stocks with the carbon tax, it's not as straightforward as stock picking associated with the supply shocks of global climate change, as discussed extensively in this book. The primary reason for the challenge with the carbon tax is that it is widely believed that (1) the carbon impact to companies will take years to phase in and (2) most of the tax to companies will likely get passed along to downstream electricity customers like you and me. Given these reasons, we will stick with the low-hanging, stock-picking fruit associated with global climate change and the associated weather-related supply shocks.

CHAPTER 10

Basic Principles of Commodity Investing

Throughout this book the focus has been to point out the specific industries and companies that will both win and lose as a result of the increasingly severe global climate changes. In this chapter we cover the basic principles involved when investing in any commodity so that you are even more prepared to take advantage of the opportunities associated with global climate change and to time your investments appropriately. (Also see Chapter 8, "Real-Life Examples: Execution, Results, and Timing," for additional insights on timing and strategy.)

First, why is it that we are focusing predominantly on commodities anyway? Well, to begin with, let's look at a specialized definition of a commodity as it specifically relates to extreme weather investing

Definition: Commodity

Bulk goods and basic materials including metals, grains, food, minerals, and energy, all derived from natural resources which may or may not trade in the futures market. The price is subject to the forces of supply and demand. Price is particularly sensitive to supply shocks associated with extreme weather events.

When looking at the definition, we can see that it references metals, grains, food, minerals, and energy, which are precisely the type of items that are directly impacted from severe weather associated with global climate change. Second, from the definition we also see that the price of commodities is affected by changes in supply and demand. One of the dominant impacts associated with global climate change is the supply shocks it causes. Because these supply shocks result in rising prices of commodities, they become directly linked to rising futures prices, stock prices, and bond prices as well—hence an opportunity for us to make some money.

So now that we understand the commodity focus, let's focus on the basic principles involved when investing in commodities. The key principles that we will cover include basic supply/demand, understanding cycles, secular rotation, timing strategies, and other basic rules of investing.

BASIC SUPPLY AND DEMAND

We have already touched on the emotions involved behind changes in supply and demand and its impact on the price of commodities, but we will add some key points to the discussion. As mentioned earlier, when investing, we ideally want to be buying commodities that fall into the Jackpot quadrant as shown in Figure 10.1.

This is the quadrant where emotions run high, which causes purchasing managers to increase the price they are willing to pay to get these commodities because they are afraid that there is not enough supply to go around. This causes prices and profits to climb for the producers of these commodities. It is common sense that buying a stock with rapidly expanding earnings is a good thing. The key question, however, is how do you know whether a particular type of commodity is in the Jackpot box or in the Big Problem box. The easy answer is to search online and get publicly available research and read the work already done by analysts on various commodities. It is generally readily available as to whether a commodity is in tight supply or in excess supply. For example, currently,

	Lot of Supply Coming	Very Little Supply Coming
Demand Growing	Neutral	Jackpot
Demand Shrinking	Big Problem	Neutral

FIGURE 10.1 Commodity Categories

copper is in tight supply and natural gas is in excess supply, both of which are widely known. Getting publicly available research is always an excellent starting point. From there, you could either go with that information if it agrees with multiple sources or you could spot check the information. It's surprisingly easy to spot check the information via Internet searches because companies love to announce publicly that they will be adding capacity/supply to the market. They do this publicly because they want to frighten away other potential competitors who may also want to add their own capacity. In addition, there are independent groups that also keep track of capacity additions in the market. Many times, multiple companies are adding capacity at the same time, even if the demand is not very strong; that's when pricing on a commodity starts to come under pressure. The more mathematically inclined reader can easily quantify these supply additions and compare them with the expected increase in demand to compare which is growing faster. The result of this analysis is what determines where the commodity falls in Figure 10.1, that is, Jackpot versus Big Problem versus Neutral.

Fortunately, we are at a point in time where many commodities simultaneously have a favorable outlook, which is based on strong global demand over the long haul but very limited available supply. Among the best of the best commodities right now are copper, corn, soybeans, metallurgical coal, and iron ore. The commodity with the worst supply/demand fundamentals right now is natural gas, as explained earlier.

UNDERSTANDING THE MATHEMATICS OF STOCK PRICES AND CYCLES

The formula for determining the price per share (P) of a company is earnings (E) times its price-to-earnings (P/E) ratio, or

$$E \times P/E = P$$

Mathematically, the "earnings" cancel out in this formula and you are left with price per share. Examples always clarify the math. I like to use DuPont (global chemical company) as an example because they never have any funny business accounting going on in their numbers and yet they are a good example of a cyclical chemical company. Take a look at their earnings, their P/E ratio, and their share price over the past decade in Table 10.1.

Before we look at this data graphically a couple of things jump out at us simply by looking at Table 10.1. We can see that DuPont's earnings hit rock bottom in both recessions, specifically in 2001 and again in 2009. This

TABLE 10.1 DuPont Earnings and P/E Ratio

DuPont	Earnings	Price/Earnings	Price
1999	2.58	25.53	65.87
2000	2.73	17.70	48.32
2001	1.19	35.72	42.51
2002	2.00	21.20	42.40
2003	1.66	27.64	45.88
2004	2.38	20.61	49.05
2005	2.34	18.16	42.49
2006	2.88	16.91	48.70
2007	3.28	13.44	44.08
2008	2.78	9.10	25.30
2009	2.03	16.59	33.68
2010	3.28	15.21	49.89
Current	3.28	16.68	54.71

Source: Public pricing data.

is classic earnings behavior. In addition, we see that DuPont has had a very impressive rebound in their current earnings, which are at a decade-long high. It's helpful to see this graphically, as shown in Figure 10.2, because it shows classic cyclical behavior.

You can see that things were great before each recession and then bottomed out in the recession, and, like clockwork, things recover again (at least for strong companies like DuPont). Impressively, the current earnings are at decade-long highs.

The P/E data are a bit harder to understand. Think of it as the price for a can of coke. I mean, really, how much are you willing to pay for each

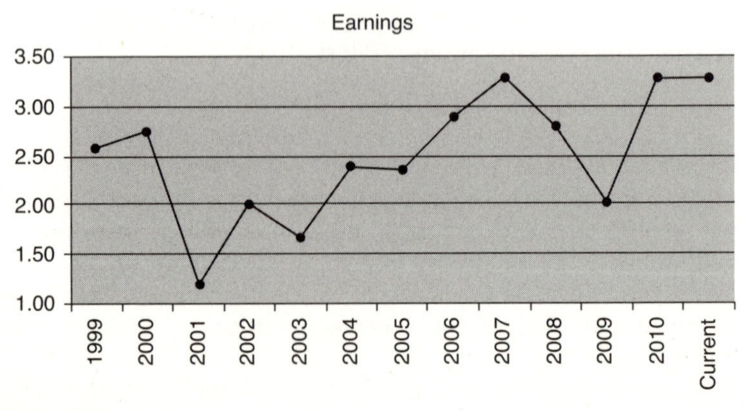

FIGURE 10.2 DuPont Earnings per Share
Source: Public filings.

FIGURE 10.3 DuPont P/E Ratio
Source: Public filings.

can of coke? In the same way the market asks itself, how much are we willing to pay for the DuPont stock for each dollar of earnings it makes? Presumably, the market is willing to pay more for each dollar of earnings if the earnings are growing very fast and less for each dollar of earnings if they are slow growers. Sadly, the market pays DuPont less today for each dollar of earnings than they used to. We can see this graphically when we put DuPont's P/E data from Table 10.1 into Figure 10.3.

As you can see, before 2004 the market was willing to pay DuPont about $25 for each dollar of earnings, but from 2005 through today the market will not go over $20 and, in fact, is willing to pay only $17 today. The problem is not just with DuPont, either. The average company in the Standard & Poor's (S&P) 500 Index is getting low-balled as well. Whenever you are investing in a commodity or any other cyclical-type stock, you have to ask yourself, "Where are we in the cycle?" The answer today is that we are in year three since the stock market hit bottom in early 2009 during the great global recession. This emergence from trough conditions has allowed the DuPont earnings to expand given the improvement in volumes and reaping the benefit from cost cutting efforts. Currently, however, the unemployment rate in the United States remains stubbornly high, the Euro region is suffering from excessive debt, and China has declared its intention to deliberately slow down its growth rate. So, despite the enjoyment of improving conditions relative to the trough, the cycle remains challenged at this point and consequently the market will not pay up in the form of a higher P/E ratio.

At some point in time, however, once the global growth rate continues its upward trend, a cyclical chemical company has the opportunity to expand its stock price in two ways. One, its earnings will rise as its sales volumes climb, and two, its P/E ratio will also climb because of a higher growth rate of earnings. When the P/E ratio increases this is called multiple expansion. The combination of higher earnings with a higher P/E ratio allows rapid expansion to its share price. To put this into perspective, if the market reverted back to the old days and was willing to give DuPont $25 for each dollar of earnings, then the stock price of DuPont would be $82 per share ($25 × $3.28 earnings = $82/share) instead of the current stock price of only $55.99 per share. This represents an additional 46 percent to the current stock price for DuPont. This is the critical driver behind secular rotation (see next section).

SECULAR ROTATION

Given the possibility of a double-whammy improvement in the stock price of a cyclical company (i.e., from both the earnings rising and the P/E rising as well), professional investors make it a habit to rotate their investments into cyclical stocks as they emerge from a recession, as they will see much bigger gains than will their investments in very defensive stocks such as food and beverage companies. This is true because the food and beverage company does not drop as far on the way down in the cycle, but likewise does not have very far to rise coming out of a cycle. This secular rotation and heavy buying into cyclical companies coming out of a recession provides additional demand and fuel for these stocks.

FUNDAMENTAL RULES TO LIVE BY WITH COMMODITY INVESTING

The following are basic rules and timing strategies to live by when investing in commodity materials in general. For additional specific strategies associated with extreme weather–based investing, see Chapter 8.

Big-Picture View

Before we drill down into the details, it is wise to understand the general big picture of how to think about these investments. Figure 10.4 defines the four basic possibilities that exist. The best of all worlds occurs when we are dealing with a commodity or a stock that is really solid fundamentally and cheap at the same time. This is the quadrant comically labeled "Back Up the Truck." We are willing to invest more into this type of opportunity. The inverse of this is a commodity or stock where the fundamentals of the

	Good Fundamentals	Bad Fundamentals
Cheap Price	Back Up the Truck	Avoid the Stock
Expensive Price	Buy Dips and Shocks	Short the Stock

FIGURE 10.4 Four Basic Investment Possibilities

business are terrible and amazingly the stock price is very expensive. For this type of situation, if we can find a suitable triggering event that will cause the price of the security to decline, we can short-sell the stock or at minimum avoid that stock like the plague.

In the case where fundamentals are terrible but the price is cheap, avoid this security. I realize it's tempting to buy on the cheap, but do not do it. It's cheap for a reason.

In the last quadrant we see good fundamentals but the stock appears rich (i.e., expensive). This is where we want to be ready to buy when the market dips, or we may also consider buying this name in the event of a global climate shock as we have discussed. This would firm up already-solid fundamentals and potentially drive the price up even further.

Is the Stock Cheap or Expensive?

When evaluating whether a stock is cheap or expensive, I like to use as many methods of valuation as possible to help confirm the result. In many ways it's like trying to decide what a house is worth. The primary trick in valuing a house is to determine how much money people have been willing spend on other similar houses in the neighborhood. After all, that's the price a willing buyer and a willing seller agreed on, so by definition it is the market price for that house. This same comparison method is used in determining the proper value of a stock. The biggest difference between houses and stocks is the number and types of features. In a house, people care about many things including how many bedrooms and bathrooms,

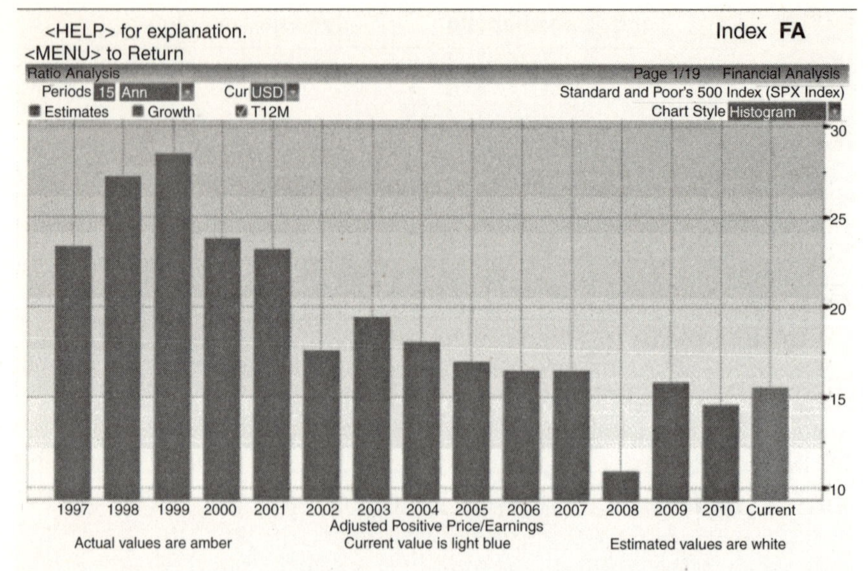

FIGURE 10.5 Price-to-Earnings Ratio for S&P 500 (Positive Earnings Only)
Source: Used with permission of Bloomberg Newswire Permissions.

acreage, proximity to schools, and on and on and on. The world of stocks, by contrast, sees things in a more straightforward manner. *The main feature investors care about in a stock is the earnings of the company and how fast they are going to grow.* So if that's true, our job is to find out how much money the market is willing to pay for these earnings, otherwise known as the price-to-earnings ratio or simply P/E. Sometimes this is also referred to as *the multiple.* The "neighbors" in stock picking can be other companies in the same industry, other companies in different industries, and historic multiples for our stock as well as the historic multiples for the market itself. Examples always help to clarify.

For starters let's look at the stock market as a whole and see how the market P/E has trended over time. Figure 10.5 shows the historic average P/E for the S&P 500 Index.

This chart is very interesting. The market is willing to pay only a little over $15 per dollar of earnings; that is, the P/E multiple for the market as a whole is only 15. Sure, it's a little higher than the 2008 recession bottom, but it's among the lowest values since 1997 and even back in 2001 during the prior recession the market was willing to pay up to $23 per dollar of earnings. So, based on this first glance, the market is appearing

TABLE 10.2 Chemical Company Price Ratio Comparison

Chemical Company	P/E	PEG	EV/EBITDA
DuPont	17x	1.5	12.6x
Monsanto	30x	1.5	15.3x
Dow	19x	2.1	10.9x
Eastman	13x	1.7	7.0x
Air Products	18x	1.6	10.1x
Median Value	**18x**	**1.6**	**10.9x**
DuPont versus Median	**6% cheaper**	**6% cheaper**	**16% more expensive**

Source: Public filings.

neutral to slightly cheap. Although this is only one quick and dirty method, it's important to get a handle on the market as a whole and not just the stock of interest because, very often, the market as a whole directionally leads the prices of many other stocks. Now let's drill down to a particular stock such as DuPont, a chemical company (ticker DD). Doing the same historic analysis, except specifically on the DuPont stock, we see the same basic pattern as with the S&P 500 as a whole. The P/E ratio is not nearly as high as it had been in or before 2004, but we can say that at least it has recovered back to prerecession levels. So, in the case of DuPont, it is fair to call the stock neutral to slightly undervalued relative to its historic multiple.

Next, we drill down again and compare DuPont to other chemical companies to see where DuPont stands relative to its chemical peers. As mentioned earlier, the more measurements, the better when it comes to valuation. As shown in Table 10.2, we did three different measurements here, the P/E, the PEG (i.e., P/E to earnings-growth ratio; after all, a stock with a higher growth rate should see a higher P/E ratio, so the PEG ratio normalizes the comparison by saying "what is the value of the P/E per unit of growth rate that this stock will see?"), and finally, the EV/EBITDA (enterprise value divided by the earnings before interest, tax, depreciation, and amortization). Some companies have so much debt that they do not have any earnings left over. In these cases, it becomes very useful to include the EV/EBITDA in the comparison. Because all of the chemical companies in this comparison have good solid positive earnings, the first two metrics are the preferred methods. So what did we learn from the table?

Based on the first two methods, the DuPont stock is 6 percent cheaper than the industry median value. I view this as neutral to slightly cheap, possibly. In addition, the EV/EBITDA method shows the DuPont stock to be 16 percent more expensive than the chemical industry median value. I view that as neutral to slightly overvalued. So if we roll together the P/E type analyses for all of the preceding situations, the basic conclusions are:

- The stock market as a whole is at the cheap end of the valuation, indicating some residual and persistent fears. Currently, the headline news fears include but are not limited to a high unemployment rate in the United States, the Chinese deliberately slowing their rate of growth, and some European governments struggling with excess debt.
- When looking at the historic DuPont P/E multiples, we get the same basic conclusion as with the market as a whole.
- Now when comparing DuPont to its investment-grade chemical company peers, we see that DuPont's valuation is somewhere in the middle of the pack and generally neutral.

This type of analysis can be repeated for any stock of interest and the same general methods will apply because the analysis of "earnings" is the common feature that ties all stocks together.

Quarterly Earning Predictions

Unless you are a professional analyst, do not get into the game of attempting to quantify the exact earnings a company will earn at the quarter press release. Instead, stick with good, solid companies with favorable long-term outlooks while also buying the companies that are not overvalued, using the P/E analysis we just covered.

Picking Market Bottoms

Do not try to pick market bottoms. When the stock market is declining as it was back in the fourth quarter of 2008 and into the first quarter of 2009, buying on the way down and then doubling the bet if it continues to decline and then, heaven forbid, tripling the bet as it further declines is a big mistake. Instead, let the market hit bottom and even bounce along the bottom for a while, and let it start turning back up again before considering getting into the market again. Yes, you may, in fact, miss out on some of the juice, but you will be far less likely to lose an enormous quantity of money on the way down. This one tip alone will help avoid a lot of unknowable guesswork.

CHAPTER 11

Opportunities in the Bond Market

Fortunately, extreme weather–based investors have ample opportunities for investing. Not only do we have the vast selection of commodities and companies in which to invest, but we also have our choice of financial markets. We can stick solely with the stock market within the boundaries of the United States if that is our comfort level or we can select stocks that are in the far corners of the Earth guided by the action plan tables throughout this book. In addition to the stock market we also have ample opportunities in the corporate bond market as well as opportunities within the futures market. In this chapter, we will talk about the basics of bond investing. We will also cover some specific opportunities within the corporate bond market. Let's begin with the purpose of owning a bond in the first place.

Why would someone even want to own a bond? The easy answer is that it is just another way for us to make some money, mostly in the form of receiving a semiannual interest payment, and also to help keep up with the rising cost of all things in our lives. Let's compare bond ownership with the old-fashioned money-under-the-mattress method to help put it into perspective. Let's assume for a minute that we are terrified of taking any risk whatsoever with our money. In fact, let's assume that the only investment that provides us comfort is to put our money under a mattress where it is safe. For the purpose of this discussion, let's temporarily forget about the fact that money under a mattress is actually losing ground because the cost of everything we buy continues to rise and yet our money under the mattress is earning zero interest. So step 1 on the risk scale in our analysis leaves us with money under the mattress earning zero interest.

Now let's move up to step 2. Step 2 on the risk scale is the investor who is tired of earning zero interest on money via the mattress method but at the same time does not really want to take any meaningful risk with his money. This type of investor might invest in government-backed bonds such as a 10-year treasury bond. In this type of investment, the investor may earn interest at the rate of about 2 percent per year (currently) and is essentially guaranteed the return of his money at the end of the 10-year term because the money is backed by a triple A (AAA)–rated government, the United States (we could also invest in a long-dated CD with our money, which also is government backed). Now the step 2 investor proudly says to himself that he has the best of both worlds. He is earning interest on his money and there is essentially zero chance of his losing it.

Let's move up to the step 3 investor. This is the lady with a slightly higher risk tolerance than the step 2 investor who was only willing to invest in government-backed bonds. The step 3 investor is willing to move into the world of corporate bonds. This is precisely the type of bonds of interest to the extreme weather–based investor. This is the type of bonds of interest because it is *corporations* that produce metals, fertilizers, and various other commodities of interest throughout this book. (There are other commodities that are produced and sold by farmers including sugar, oranges, and many others, which are not produced by corporations. For these types of investments, we make use of the futures or exchange-traded fund (ETF) markets. See Chapter 13, "Basic Principles of Futures Market Investing"). Now let's dig down and understand the key points of investing in corporate bonds.

OK, let's get back to the step 3 investor. If I, as a step 3 corporate bond investor, am willing to buy corporate bonds that are not backed by the U.S. government, then I better be earning interest on this investment that is even higher than the interest I could earn on a 10-year Treasury bond. After all, with higher risk comes higher reward. This is the fundamental, expected truth behind all investments. The key question is how much more interest? (This additional interest on top of the interest we would get for simply buying a government Treasury bond is called the "spread.") Well, it depends on the corporation that issued the bonds. It makes sense that if I buy the bonds of a very solid company with very low debt levels (and hence very low risk of bankruptcy) and a business profile that is highly profitable and very stable year in and year out, then I should receive interest on these bonds that is only slightly higher than the interest I could earn on a government-backed 10-year Treasury bond because this corporation has relatively little risk due to its stable nature and low debt load.

However, if I am willing to be a lot riskier and buy the bonds of a company that has plenty of debt (and therefore a higher risk of bankruptcy), low profit margins, and a generally unstable business profile, I would expect to receive a much, much higher interest rate. After all, my risk and reward should go hand in hand.

Trying to determine the appropriate interest we should receive on a corporate bond is kind of like trying to determine the appropriate price on a new house. How do we know what the appropriate price on a house should be? We look at very similar houses and see what people were willing to pay for them in recent actual transactions. We do precisely that in the corporate bond market as well. We simply look at other similar corporate bonds (i.e., "houses") that are in the market to find out a fair price. In some ways, comparing corporate bonds is easier than comparing houses because there are fewer critical features in a bond then there are in a house. The simplest and broadest feature of a corporate bond is the corporate credit rating. Lower credit ratings command higher interest on your investment. Fortunately, the three primary credit agencies including Moody's, Standard & Poor's (S&P), and Fitch, provide this information to the investing public. Table 11.1 shows Moody's ratings scale from the best possible at Aaa (i.e., the U.S. Government has this rating) down to a rating of D, which stands for a company that is in default (i.e., didn't pay their interest payment).

As shown in Table 11.1, investment-grade companies have a rating of Baa3 or better. Likewise, below-investment-grade companies, otherwise known as *high-yield bonds* or *junk bonds*, have ratings of Ba1 or lower. The rating agencies come up with ratings by looking at various financial metrics and other characteristics of the company. For example, higher-rated companies will have a higher level of free cash flow as a percentage of their total debt load, among other things.

So, as an investor of corporate bonds, we have a basic choice of whether to go with investment-grade companies or to get a bit riskier in order to get higher returns and go with below-investment-grade companies. Even within these two categories we have additional choices to make, including what type of investment-grade company we want to invest in, for example, a Baa3 or an A3. Again, even within the investment-grade category, lower ratings command higher interest on the investment. Similarly, if we choose to go with a below-investment-grade, high-yield company, we have a choice to make. We can go with a Ba2 company or a B2 company, for example. A double-B-type company, in general, is far less risky than a single-B company. Whatever the rating of the company we choose to go with (i.e., corporate bonds we choose to buy), then that is the rating we use for price comparison (remember the house comparison example). Before digging down into actual corporate bond examples, it would be helpful to look at the bigger picture in terms of bond investing, as depicted in Figure 11.1.

As shown in Figure 11.1, the strong preference when buying a bond is good strong company fundamentals. Recall from Chapter 1 how we split

TABLE 11.1 Moody's Corporate Credit Rating Scale

Moody's Rating Scale
Investment Grade:
Aaa
Aa1
Aa2
Aa3
A1
A2
A3
Baa1
Baa2
Baa3
Below Investment Grade (or "High-Yield" or "Junk" Bonds):
Ba1
Ba2
Ba3
B1
B2
B3
Caa1
Caa2
Caa3
Ca1
Ca2
Ca3
C1
C2
C3
D (company in "default")

Source: Public filings.

up the various commodities into either the Jackpot category, the Neutral category, or the Big Problem category. When we are talking about "good fundamentals" in Figure 11.1, we are generally referring to the Jackpot commodities but also occasionally to the Neutral commodities.

On the other axis is the bond price (i.e., "house price") comparison against other similar bonds. If we are evaluating a target bond with a rating of Baa3 and it will be offering us a return of 5 percent and we compare this to other similarly rated bonds that, on average, provide a return of only 4 percent, then we would say that this is attractive relative to the current market pricing. Another way to say this is that the 5 percent bond has a wider spread over a similar time length government Treasury bond of, say,

	Good Fundamentals	Bad Fundamentals
Wide Spreads	Back Up the Truck	Avoid the Bond
Tight Spreads	Buy Dips and Shocks	Short or Avoid the Bond

FIGURE 11.1 Four Basic Investment Possibilities

3 percent. In this case the spread over Treasuries is 2 percent (5 percent − 3 percent = 2 percent, or 200 basis points is another way to say the same thing because 1 percent = 100 basis points), whereas the average spread of other bonds with similar ratings over Treasuries is only 1 percent (or 100 basis points), which we get from taking the 4 percent return on the average corporate bond with this Baa3 rating at the time of your analysis and the 3 percent you could get if you were only willing to invest in the riskless government Treasury bond. We like to have wider spreads for our target bond as compared with the average spread for a collection of similarly rated bonds in the market at any point in time, as long as the strong fundamental story for the company remains in tact.

Likewise, if our target bond could provide us a return of only 3.5 percent but the average return for similar bonds in the market with essentially the same rating was even higher, at 4 percent, then we would say our target bond is trading with "tight" spreads. After all, our target bond in this example would have a spread over the 3 percent Treasury bond of only 0.5 percent (3.5 percent − 3.0 percent = 0.5 percent, or 50 basis points), compared with the average bond with similar rating in the market with a 1 percent spread (4.0 percent − 3.0 percent = 1.0 percent, or 100 basis points).

So taking the two axes of the grid into account, the best possible bond to buy at any point in time is the combination of good fundamentals with wide spreads. That combination earns the label "Back Up the Truck," a comical way of saying you are getting a good return on this bond for its level of risk as defined by its corporate rating. This quadrant is also attractive when the action plan tables that are covered extensively in this book say to buy the stock or bonds of a biggest winner company.

The opposite end of the scale is having bad fundamentals for a particular company or commodity and yet having tight spreads. This means that the market is not paying you to take on this extra risk. Generally speaking, we "Avoid the Bond." We also said in the table that this bond could actually be a candidate for shorting, a technique for betting against this bond as opposed to buying this bond (see Chapter 9, "Playing Both Sides of the Coin," for additional information on shorting). In the bond market, shorting a position is often left to the professional investor. For our purposes, we would simply not buy this bond.

What about the other two quadrants of Figure 11.1? Well, in the cross-section of good fundamentals with tight spreads, we have a company that falls into the Jackpot category, meaning the fundamentals are great and, guess what, the market already knows that. The reason we know the market already knows that is that spreads are tight, meaning everybody wants to buy the same bond (we will get into the numbers behind the bonds shortly, where we will learn that the price of the bond moves in the opposite direction of the yield on the bond). Sometimes, we can get away with buying this bond anyway, even though spreads are tight. We can do this if two conditions are met: (1) we are certain that the fundamentals of the bond are good (i.e., Jackpot category); and (2) the return we would get on this bond is acceptable to us, for example, 3.5 percent. If you are happy with getting that return for your money, then go ahead and buy the bond. This is the type of bond where we can also be patient and wait for dips in the market when prices of all securities slip a bit, causing spreads to widen, thus making the bond more attractive. Buying the bond when the market prices dip is known as "Buying on the Dips." This quadrant is also attractive when the action plan tables that are covered extensively in this book say to buy the stock or bonds of a biggest winner company.

The final quadrant in the Figure 11.1 is the cross-section of bad fundamentals with wide spreads. This is an interesting combination because it says that, yes, you are dealing with a weak company, and the market is willing to pay you for it with wide spreads. Do not get hooked like a fish on wide spreads as payment for buying a terrible company. It sounds alluring, but don't do it. The extreme weather–based investor prefers to stick with bonds that fall into the Jackpot category or even occasionally into the Neutral category, but, generally speaking, we like to avoid bad companies regardless of how wide the spreads get.

You should now have a good understanding of the basics of bond investing. We accomplished this without having to dig into the math behind the qualitative factors. So if you have no interest in the math behind the qualitative factors, then feel free to skip the remainder of this section. We will also be talking about various actual bonds that are in the market today, which will further aid in your understanding.

Before we look at some real-life examples, let's talk about where corporate bonds come from. Corporate bonds are nothing more than debt to a company. A company borrows money for a host of reasons. Maybe they want to acquire another company. Maybe they want to expand a manufacturing plant. Maybe they want to borrow money so they can pay a dividend to the stockholders. Maybe they simply want to refinance another outstanding bond that is maturing. Let's say a company needs to borrow $500 million. The first thing it does is hire a couple of investment banks, let's say two, which each ultimately give the company $250 million. So now the company has its $500 million, and they go away happily using the money for their purposes. The investment banks hold basically a contract, with the company promising to pay the debt back in say 10 years (which could be shorter or longer than 10 years depending on the desired maturity of the company) with semiannual interest payments, which are also known as the coupon payments. The investment banks do not actually want to hang on to this contract, otherwise known as corporate bonds. Instead, they allocate them or sell these bonds to potentially hundreds of other bond-buying companies like hedge funds or insurance companies. Let's say in total there are 50 smaller investment companies that buy the $500 million worth of these corporate bonds, so the average investment fund would get allocated $10 million worth of these bonds.

Now we finally have a market for these bonds where institutional investors and retail investors like you can go out and buy and sell these bonds (retail investors can go through their existing stock trading company like Ameritrade to buy bonds as they would with stock purchases). Similar to publicly traded stock, as these bonds are bought and sold, the company that issued this debt does not receive a dime of more money. They already got their $500 million in step 1 and therefore they are no longer involved in the secondary market trading that takes place at this time.

OK, now that you understand the origin of a corporate bond, let's take a look at some real-life examples. For starters, we will look at an investment-grade corporate bond. We will start with Teck Resources, as shown in Table 11.2.

Interestingly, the Teck Resources bond shown in Table 11.2 is just one of many corporate bonds available for Teck Resources. The reason there are many available is because every time Teck Resources needs to borrow money, they come back to the bond market and issue more bonds; meanwhile, they very often still have their other bonds outstanding. As shown in this table, this bond matures in the year 2021. That is the year they will need to pay back the debt associated with this particular bond. At that time, they may refinance this bond with a new bond or they may simply pay off

TABLE 11.2 Teck Resources Corporate Bond

Company	**Teck Resources**
Major products	Metallurgical coal and copper
Size	$500 million
Rating	Baa2
Maturity	2021
Coupon (interest payment)	4.5% annual interest payment (half paid every 6 months)
Price	103.3
Spread to Treasuries	1.17% or 117 basis points
Yield to worst (YTW)	**4.10%**

Source: Public pricing data.

this debt with cash if they have enough cash at the time. So, as you can see, the particular corporate bonds available for a company changes regularly. However, the stock for the company remains the same. So at any one time, we can compare the bond in question with not only other bonds from other companies available in the market with similar ratings but we can also compare all of Teck Resources' bonds against each other to see which one is the most attractive.

Now let's talk about the features of this particular bond. As shown at the time of the issuance of this bond, Teck borrowed a total of $500 million. The rating on this bond is investment grade at Baa2 and has a maturity of 2021. In addition, this bond has a coupon (interest) payment of 4.5 percent per year, half of which is paid two times per year.

The next three features, including the dollar price, the yield to worst (YTW), and the spread to Treasuries require more explanation.

Let's start with the dollar price. Bonds generally start with a dollar price of 100 (i.e., 100 cents on the dollar), but as time goes on the dollar price changes just like the stock price of a company changes over time. Technically speaking, the bond's price is the present value of all future cash flows associated with this bond, including two specific future cash flows: (1) the coupon (or interest) payment that we receive every six months and (2) the final payback of our original dollar principal amount at the time of maturity of the bond or in 2021 in this Teck bond. What do we mean by present value? The present value concept is related to the "time value of money." Unless we are fans of putting our money under a mattress, we like to earn interest on our money. After all, if I put a $100 bill under my mattress for a year, it will still be a $100 bill a year later. If I, however, earn 10 percent interest on my money by investing it, then I would have $110 at the end of the year. The "present value" and "time value of money" concepts are referring to the interest rate we earn on our money—nothing else. Using the $100 bill earning 10 percent example, we can show how we end up getting to $110 by the end of the year in this way:

$$\$110 = \$100 \times (1 + \text{IR})$$

Another way to write this is shown below, since we know in this example that the IR (interest rate) is 10 percent, which can also be written 0.10.

$$\$110 = \$100 \times (1 + 0.10)$$

So, as you can see, the $100 we have right now is called the present value and the $110 is called the future value. So if we were to flip-flop the numbers, we can say the same exact thing but going in reverse:

$$\text{Present Value} = \$110/(1 + 0.10) = \$100$$

So when we go in reverse, starting with the future value, we take the future cash flow numbers and divide them by the interest rate instead of multiplying them as we did when calculating future values. This simple method of discounting future values to obtain the present value is precisely how we determine the present value (or dollar price) of a bond.

Now let's take it one step further. Bonds do not just pay interest for one year; they pay for many years in a row. As a matter of fact, they pay two times per year for many years in a row. We will use a five-year bond for our example with a 10 percent coupon (annual interest rate). To make the math simple we will assume they pay only one time per year. So, let's assume that this particular bond is priced on day one at 100 cents on the dollar, as is often the case. For every $100 of bond we own, we would get $10 per year in interest, as discussed earlier, and at the end of year five we would also get back our original investing principal of $100. At this point, we know three things:

1. The starting price (i.e., the present value of future cash flows) at 100.
2. The annual interest payment of 10 ($100 × 10 percent/year = 10).
3. The investing principal amount of $100 is returned to us at the end of year 5.

So, given these three things, the present value (PV) calculation is shown below. Notice it is simply doing what we did earlier, except we have to do it for each individual cash flow into the future. Also notice the superscript numbers. In year three, for example, we have to discount the cash flow back three times, once for each year, hence the three. In year five, notice we discount back 110, which includes not only the 10 interest payments in year five but also the original 100 we get back from the company as the bond matures in year five.

$$PV = 100 = 10/(1 + IR) + 10/(1 + IR)^2 + 10/(1 + IR)^3 + 10/(1 + IR)^4 + 110/(1 + IR)^5$$

The IR (interest rate) that we use to discount back to our known present value is actually called the yield to maturity (YTM). Up until this point we have been calling it the interest rate generically to simplify things. Mathematically, when the price of the bond is at 100 (i.e., its present value), the YTM and the coupon rate are identical. So, in this example, because the price of the bond is initially 100, then the YTM is 10 percent and the coupon payment we receive is also 10 percent. This is where the math becomes graphical to help us understand the relationship between a bond's price and a bond's YTM. Without even looking at the graph yet, we can guess that there is an inverse relationship between these two things because as you can see from the present value formula, the price of the bond is related to 1 divided by the YTM (i.e., the inverse of the YTM). So, because of that relationship, and because common sense tells us that we get a smaller number answer when we use a higher discount rate, we can expect the graph to show that as the yield to maturity goes higher, the price of the bond will go lower. The graph of that larger equation is shown in Figure 11.2.

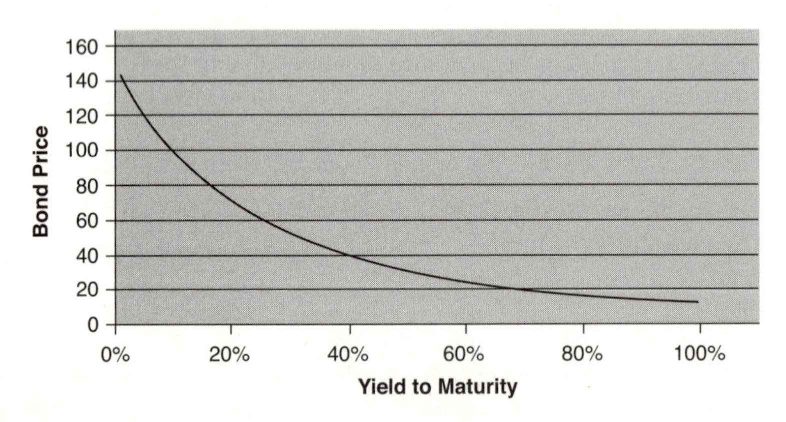

FIGURE 11.2 Bond Price versus Yield to Maturity

A picture tells a thousand words. From this simple graph, we can learn an awful lot about bond investing. The major lessons we can learn from this graph are the following:

- Recall that in this particular example, we, the bondholder and bond investor, are receiving a coupon (or interest) payment of 10 percent per year. Notice that when the bond price equals exactly 100 cents on the dollar, the YTM also exactly equals the coupon rate, 10 percent. This is not a coincidence; it's a mathematical certainty, no matter what the coupon rate.
- Directionally, we can clearly see that bond prices and yields go in opposite directions. When the yield to maturity goes higher, bond prices go lower, and when yields go lower, bond prices go higher.
- So, as a bond investor, we like it when interest rates and inflation in general go lower because mathematically the bond price goes higher. The old investing truism "buy low and sell high" also applies to bonds.
- You may have noticed that the curve is bent, having what is called convexity. Notice that as yields decline lower and lower, the bond price climbs higher and higher, and this occurs at a faster and faster rate. Similarly, as yields rise, bond prices fall but at a slower and slower rate as the curve flattens out. Before we get too excited about rates plummeting and bond prices skyrocketing, we have to realize that the company hates paying a 10 percent coupon when market interest rates are at only 4 percent, for example. So we, as the bondholder, run the risk that the company may very likely refinance their bond with a bond that has a cheaper coupon rate payment. Technically speaking, the bond price may never get as high as shown in the graph at the very low interest rates.
- Because of the potential for the company to call the bond before maturity, the yield metric of most interest to the bond investor is the yield to worst (YTW). The YTM (i.e., yield to maturity) might be absolutely meaningless if the company plans on taking the bond out long before maturity (i.e., "calling" the bond). We could also ask ourselves what is the yield to call (YTC) if we are worried about the company calling the bond before maturity, but in general the YTW is the best single metric because it tells us the worst case scenario in terms of how much return we get regardless of the company's action as long as the bond gets taken out eventually (at maturity or earlier).

Now that we understand the fundamental mathematics of bond investing, let's have another look at the actual Teck Resources bond discussed earlier (see Table 11.2).

As you can see, the YTW is 4.1 percent on this bond. This means whether the company takes this bond out at maturity or calls it earlier, the worst the bond investor will do is to see a return on his money of 4.1 percent. This, of course, assumes that the bond is eventually taken out, and it also assumes that the company does not go bankrupt along the way, hence our primary criterion as weather-based investors of investing in companies that are solid fundamentally and with conservative balance sheets (i.e., not too much debt for the size of their company).

Notice also that because the dollar price of the bond is at 103.3, and is higher than 100, that the YTW at 4.1 percent is lower than the coupon percent, which is at 4.5 percent. This is not a surprise as prices move in the opposite direction as yields, as we learned from the graph. If the dollar price were exactly 100, the yield would match the coupon rate at 4.5 percent.

The final metric that we watch is the spread to Treasuries. Recall from our earlier discussion that the extra interest we receive on our corporate bond, over and above the interest we would earn if we simply bought a similar maturity Treasury Bond, is the "spread." Notice in the Teck Resources bond, the spread in this case is 1.17 percent or 117 basis points.

Now let's compare Teck Resources with some other companies that have a similar rating to see where Teck falls in terms of the evaluation grid. Specifically, let's evaluate Teck to see where it falls on both axes, including its fundamentals and its pricing. On the fundamental side, we can see from Table 11.3 and from the discussion in Chapter 1 that both metallurgical coal and copper are in the Jackpot commodity category at this point in time. So, for starters, that is a huge positive.

What about pricing? Let's take a look at Table 11.4 comparing Teck's bonds to the bonds of its similarly rated peers.

The basic conclusions from Table 11.4 are the following:

- The YTW for Teck is in the middle of the pack, which is reasonably priced.

TABLE 11.3 Company Comparisons

Company	Key Products	Commodity Category
Teck Resources	Metallurgical coal and copper	Jackpot
Alcoa	Aluminum metal	Neutral
Arcelormittal	Steel	Neutral
Dow Chemical	Chemicals	Jackpot

Source: Estimates.

TABLE 11.4 Corporate Bond Price Comparisons

Company	Coupon	Dollar Price	Bond Maturity	Rating	Yield to Worst (YTW)
Teck Resources	4.5%	103.3	2021	Baa2	4.1%
Alcoa	5.4%	102.3	2021	Baa3	5.1%
Arcelormittal	5.5%	100.8	2021	Baa3	5.4%
Dow Chemical	6.0%	113.9	2017	Baa3	3.5%

Source: Public pricing data.

- Teck has the best rating by a small margin and therefore would expect to see slightly tighter (lower) yields as a result.
- The Dow Chemical bonds have a maturity date that is four years before Teck, partly explaining why the Dow bonds have a lower yield. In general, longer-dated maturities require higher yields to compensate for holding the bonds for the extra years and the associated risk.
- The coupon at Dow Chemical is fairly high, at 6 percent, so the dollar price of the bonds is much higher, which brings the YTW back down. The market is saying you will have to pay up in order to get that higher coupon. The net effect of coupon and dollar price is a yield at the Dow bonds of 3.5 percent.
- The dollar price of Teck is slightly above par (i.e., 100) but even with a higher than par dollar price, the bonds are still yielding 4.1 percent, which is reasonable relative to the other names in the table.

Remember that each company can and often does have more than one bond outstanding and so the investor can and should compare against not only various other similarly rated companies, but also against other bonds within the same company in order to find the best possible combination of yield to worst and company fundamentals.

To summarize for Teck, the company's primary lines of business currently fall into the Jackpot category. This is good. On the pricing side, the bonds are generally fairly priced (i.e., its yield to worst) relative to its peer group. The combination of these two factors makes Teck a reasonable bond investment, not just for an extreme weather event but for ownership in general.

So, in relation to global climate shocks and extreme weather events, how do we put this newly found corporate bond knowledge to work? Throughout this book, numerous action plan tables have been created. Each action plan was specifically created to deal with specific global climate shocks and extreme weather events. The body of each action plan *ranks* from best to worst not only the biggest winners but also the biggest losers in every global climate scenario. So when you see a company at the top of the biggest winners list, you have a choice, depending on your comfort level as an investor. Assuming that the particular company at the top of the biggest winners list has both publicly traded stocks and bonds, you can buy the stock, the bonds, or both. Recall from our earlier discussion, however, that a company has only one stock but very often many different bonds available from which to choose. Because the bonds of a company get refinanced or paid off over time, we will not list out the current bonds for each company because they change quite frequently. If you choose to buy the bonds of a biggest winner company, then simply apply the basic analysis shown in this chapter to help you select the best bond available.

OPPORTUNITIES IN THE MUNICIPAL BOND MARKET

The municipal bond market is a unique and very interesting case for the extreme weather–based investor. Generally speaking, when a global climate shock or an extreme weather event takes place somewhere in the world, it negatively impacts the region that was hit, but it simultaneously helps other regions of the world that were not hit. For example, if there is a severe drought in Russia (which actually occurred recently), this will act as a supply shock to the wheat market. This obviously hurts the Russian farmers. However, the supply shock in the global wheat market causes the price of wheat to go up for all of the remaining non-Russian wheat farmers around the globe that still have wheat to sell.

This basic model, which works very effectively, as shown repeatedly throughout this book, does not generally work in the municipal bond market. The reason it does not work is due to the lack of a strong correlation between municipal bond regions in relation to extreme weather occurrences. An extreme weather event, such as Hurricane Katrina hitting New Orleans, causes the municipal bonds in that region to not only see rating downgrades but also to experience bond price weakness and volatility. However, just because the municipal bonds in New Orleans (in this example) have been hurt does not mean that municipal bonds in a west coast state benefit. There is little upside opportunity with regard to extreme weather–based investing in municipal bonds.

Therefore, given this interesting response of municipal bonds to extreme weather events, our action plan with regard to this market is to avoid buying the municipal bonds in the regions that are expected to see extreme weather events. How can we translate this directive into practice? We adhere to the following:

- This is one area where we can benefit from historic patterns. Specifically, do not buy excessively in the municipal bond market in hurricane regions during hurricane season.
- Do not buy excessively in the municipal bond market in regions where it is well known that heavy snow and potential associated cost overruns could impact municipal bond pricing.
- *In general, our action plan in the municipal bond market is one of avoiding potential pitfalls as opposed to searching for upside opportunities, as was the case for many other sectors in this book.*
- See Chapter 8 in the section titled, "The Rules of Extreme Weather-Based Investing," where we cover an advanced technique in extreme weather-based investing where it is occasionally acceptable to buy "the loser" under certain conditions.

CHAPTER 12

Opportunities in the Foreign Currency Exchange Market

As surprising is it may sound, the foreign exchange market is far larger than even the stock market. It therefore behooves us to give careful consideration to the potential investment opportunities in this market in relation to global climate shocks or extreme weather events. *As will be demonstrated in this chapter, the basic final conclusions are as follows:*

- Due to the massive amount of commodity production in numerous countries around the world, many countries have currencies that have become known as "commodity currencies." In particular, a large percentage of their total exports are from commodities that they have produced, and therefore the value of their currency is generally correlated to the price of the relevant commodities that they produce. This is a well-known phenomenon. The fact that there are commodity currencies implies potential investment opportunities in relation to global climate shocks and extreme weather events because such events generally benefit commodity prices and therefore will also benefit commodity currencies.
- In our analysis we focus on the commodity currencies from Canada and Australia primarily because these currencies are readily traded in the foreign exchange markets and both countries are politically stable. What we find is that both of these countries produce a wide variety of commodities. Unfortunately, extreme weather events do not necessarily impact all commodities types at one time. In fact, very often, an extreme weather event will impact only a single commodity or possibly

two commodities at any one time. Although it is true that a price spike in one of the key commodities of a country helps strengthen their currency, the effect is highly diluted because the commodity is only one of many commodities in their export product mix. As a result, the impact of an extreme weather event has only a secondary, indirect benefit to the value of the country's currency and therefore better, extreme weather–based investment opportunities exist in other noncurrency-type investments, as discussed extensively throughout this book.

- When dealing with global climate shock or extreme weather–based investing in currencies, some effects are counterintuitive. Take, for example, the recent massive flooding that covered the eastern regions of Australia. The flooding caused an extended interruption to coal exports out of Australia. Because Australia is a major global player in seaborne coal, the global price of coal shot upward. However, rising coal prices do not increase the value of Australian exports *(and therefore do not increase the value of the Australian currency)* if Australia is not able to export coal out of flooded coal mines! Meanwhile, because the global price of coal shot upward, other commodity currencies such as the Canadian dollar can benefit because they, too, export coal and are not experiencing a flood! So we must be very careful. Just because the global coal price is rising does not mean that the Australian dollar is strengthening in particular when Australian floods have blocked coal exports. In fact, if we look at the one-month time frame from November 1, 2010, until November 30, 2010, which corresponded to the ramping up of the flooding situation in Australia we see that the Australian currency actually declined in value by 3 percent but the Canadian dollar declined in value by 1 percent, thus outperforming the Australian dollar by 2 percentage points over this one-month time frame. The Australian dollar did not benefit from rising coal prices, as expected. Even in the case of the Canadian currency, despite the fact that they benefited from rising coal prices, their currency also did not appreciate. This is because the positive effect of rising coal prices was diluted by the numerous other commodities coming out of the region. *The critical point here is that in the event of a global climate shock or extreme weather event that affects a particular region, the best currency investment is to avoid buying that particular commodity currency of the country that is seeing impact from the extreme weather event.* This investment *avoidance* is the preferred and most conservative play, as compared with buying the commodity currency of other regions that export the same commodity but are not seeing the global climate shock. This is because of the dilutive nature of Canada (in this example) and its vast quantity of commodity types coming out of that region.

- In general, we identify in this chapter two types of commodity currencies. Class 1 commodity currencies (i.e., Canadian and Australian dollars) are acceptable from the standpoint of being politically stable and having adequate liquidity but fail because they suffer from commodity dilution. Class 2 commodity currencies (i.e., various African-based currencies) satisfy the problem of commodity dilution but fail because they are politically less stable and often have inadequate liquidity. In both cases the conclusions are the same for an extreme weather–based investor. The conclusions are twofold: (1) avoid buying the currency for the country that was hit by the extreme weather event; and (2) instead, stick with buying the "biggest winner" companies and commodities identified in the action plan tables throughout this book.

Now that we have covered the fundamental final conclusions relating extreme weather–based investing with the foreign exchange market, we will drill down into some of the detail behind the conclusions.

CANADIAN DOLLAR

The value of the Canadian dollar has strengthened greatly over the years. We can see this graphically over the past decade in Figure 12.1.

A point of clarity on the graph in Figure 12.1, notice that the units in this graph are number of Canadian dollars per one U.S. dollar (i.e., $Can/$US). Do not be confused by this format. A simple analogy clears up this meaning.

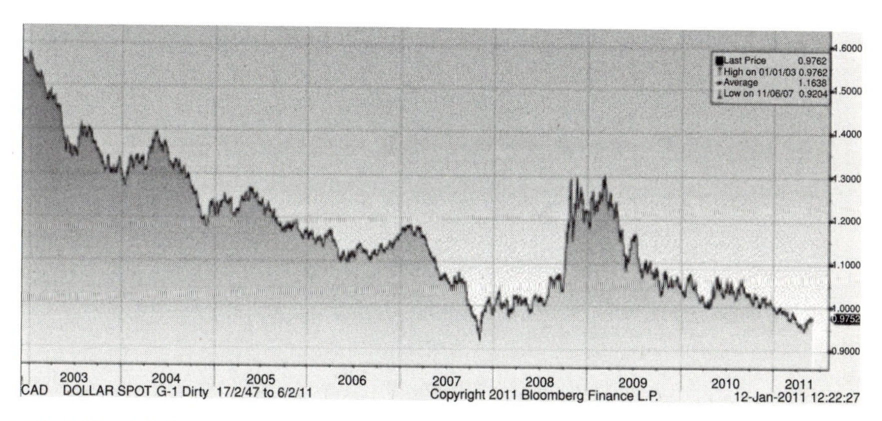

FIGURE 12.1 Value of the Canadian Dollar, $Can/$US, 2002 to 2011 (Up 62 Percent)
Source: Used with permission of Bloomberg Newswire Permissions.

What if we instead said the number of Canadian dollars for a can of soda. Then the graph with the declining slope would mean that the can of soda (or U.S. dollar in this case) was getting cheaper over time. So if the U.S. dollar is getting cheaper over time, then that must mean the inverse for the Canadian dollar. Over the past decade the value of the Canadian dollar has strengthened by 62 percent, as shown in Figure 12.1. A key question is why has the Canadian dollar shot up in value so greatly over the past decade?

Many people like to refer to the Canadian dollar as not just a commodity currency but more specifically as an oil currency. This would be an acceptable term if oil were the only commodity exported out of the region. However, as shown in the pie graph in Figure 12.2, Canada exports many different types of commodities, including oil, natural gas, fertilizer, metals, timber products, and various commodity chemicals. None of these commodities hold a dominant position in Canada's exports, as shown in Figure 12.2, and therefore it is not entirely accurate to refer to this currency as strictly an oil currency, although the price of oil does filter into the price of other commodities because it is an energy input cost, so the "oil currency" label is partially accurate.

Interestingly, 56 percent of the exports coming out of Canada are commodities. Given that the majority of exports are commodities, the label "commodity currency" is much more accurate than to say "oil currency."

In fact, if we look at a basket of eight major commodities in Table 12.1, all of which are produced in Canada, and create a correlation table covering the prior three-year time period, we obtain the following matrix of cross-correlations.

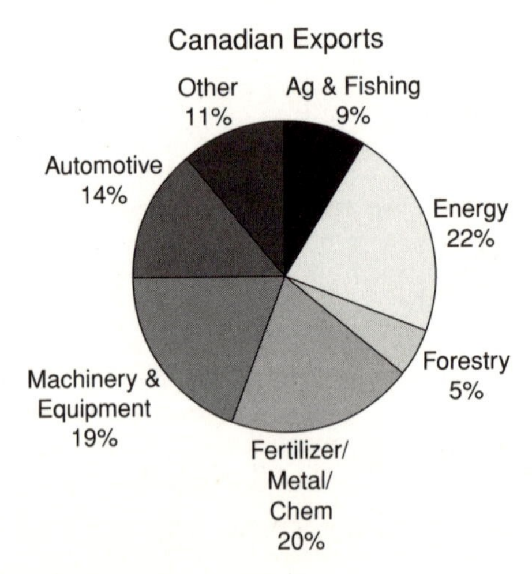

FIGURE 12.2 Canadian Exports
Source: Canadian government statistics, 2010.

TABLE 12.1 Correlation Table among Canadian Commodities

	Copper	Nickel	Gold	Silver	Natural Gas	Aluminum	Oil	Corn
Copper	100%							
Nickel	94%	100%						
Gold	86%	85%	100%					
Silver	83%	79%	92%	100%				
Natural Gas	-11%	-21%	-30%	-26%	100%			
Aluminum	91%	86%	68%	68%	24%	100%		
Oil	85%	78%	58%	65%	23%	94%	100%	
Corn	64%	51%	67%	84%	11%	64%	62%	100%

Source: Public pricing data, 2008 to 2011.

What can we glean from Table 12.1? It becomes immediately obvious that there is a very strong, positive correlation between each and every commodity combination with the exception of natural gas. As discussed earlier, natural gas falls into the Big Problem category because not only was demand soft in the North American region over this time frame but the supply side was in excess due to the shale gas–type discoveries in the United States. There is also one other exception. Five percent of the pie graph of exports coming out of Canada is labeled "Forestry." A fraction of this 5 percent of "Forestry" refers to lumber. Of course, the lumber market has suffered over this time frame due to the weak housing market, particularly within the United States. However, aside from the softer natural gas and lumber markets, the great majority of the commodities exported out of Canada have enjoyed a highly correlated rising price environment. It has been the combination of bull market pricing (i.e., prices that go up) across the entire commodity basket, which has greatly contributed to the rising value of the Canadian dollar over the past decade.

So when it comes to extreme weather–based investing, what have we learned regarding the "commodity currency" in Canada? We've learned a couple of things:

1. Because weather-based investing is often linked to commodities, it implies some level of investing potential within the "commodity currency" markets. However, extreme weather events very often only impact one or two commodities at a single point in time, and therefore the commodity price spike associated with the global climate shock gets very diluted in Canada because of the vast quantities of commodity types coming out of the region, thus negating the effectiveness of a Canadian currency investment.

2. Given point one, the more effective extreme weather or global climate shock–type investment is to go after the specific, affected commodity and its associated stocks, bonds, and futures.

AUSTRALIAN DOLLAR

The Australian dollar, similar to the Canadian dollar, has increased in value over the past decade. The Canadian dollar increased 62 percent over the past decade. The Australian dollar, by contrast, has appreciated a whopping 88 percent over the same time frame. If we look at the basket of exports coming out of Australia, we obtain the pie chart shown in Figure 12.3.

As shown, iron ore plus coal sum to 40 percent of the total exports coming out of Australia. Recall from Chapter 1 that iron ore has a Jackpot story, and so does coal in general in the region, particularly metallurgical coal. So, similar to the discussion on the commodity currency in Canada, the commodity currency in Australia has also enjoyed a very bullish basket of commodity pricing, thus helping to appreciate this currency by 88 percent over the last decade.

The key differentiator between the two currencies, however, is the key trading partner. In the case of Canada, the dominant trading partner is the United States. In the case of Australia, the dominant trading partner is China. The United States took part in the global recession with a recession of its own. China, by contrast, did not participate in the global recession.

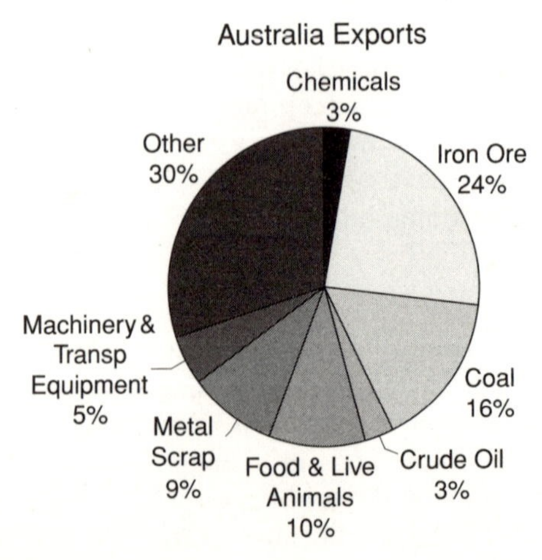

FIGURE 12.3 Australian Exports
Source: Australian Government Bureau of Statistics, 2010.

In the depths of the global recession, the gross domestic product (GDP) in China dropped to a roaring and positive 6.5 percent. So not only does Australia get to enjoy commodity prices that have been rising over the past decade, but they also get to enjoy much stronger volume growth rates than their Canadian peer group who feeds the relatively slow growth and mature United States.

Despite its superior position given its close proximity to China, even Australia cannot escape extreme weather and global climate shocks. Beginning in late 2010, very heavy rains and flooding began to accumulate in the eastern portion of the country. This effectively flooded the coal mines, which are geographically rich in quantity in eastern Australia. Recall from above that coal represents 16 percent of exports out of Australia, a meaningful quantity but still heavily diluted among various other commodities coming out of the country. Fortunately for Australia, the iron ore mines are in the western Australia region and were not affected. Interestingly, both Australia and Canada produce and export coal. So, given the supply shock in Australia, one would expect that the commodity currency in Canada would outperform the commodity currency in Australia because it enjoyed rising coal prices and avoided the floods As we already learned from our earlier discussion, however, even the Canadian currency did not appreciate as a result of the dilution effect as discussed above.

The Australian currency by contrast declined as expected because 16 percent of its exports were cut off. However, the dilution effect in this case actually helped to stem the decline in the value of this currency. The bottom line guidance for the extreme weather-based investor is to avoid buying the currency of the country that is taking a direct hit from an extreme weather event. This currency-avoidance strategy is preferred over outright shorting of this currency because of the dilution effect, which results from the wide variety of commodities produced in the region.

OTHER COMMODITY CURRENCIES

The reason Canada and Australia were at the center of our commodity currency focus is that these investments are highly liquid (plenty of opportunity to readily buy and sell when desired) and they are both countries that are politically stable. Both of these factors are prerequisites for an extreme weather–based investor.

There are other commodity currencies that meet these two criteria such as Brazil and Chile. However, they also fall victim to the dilution effect mentioned earlier. In both cases, although commodities are a large percentage of exports, both countries have a diverse variety of commodity types, thus negating the upside potential in the event of a global climate shock in a region outside of Brazil or Chile. *Once again, the action plan for an extreme weather–based investor within the currency market is to simply avoid buying the currency of the country seeing the global climate event.*

What about countries that do not suffer from the dilution effect? That is, what about countries that export only one major commodity? Many countries within the African region meet this criterion but then lose out because they do not meet our fundamental prerequisites of having good liquidity and being politically stable. An example of such a currency is in the region known as Cote d'Ivoire, where the currency is the CFA franc BCEAO (northwestern Africa). This region is responsible for nearly 40 percent of global cocoa bean production. In the event of a global climate shock or extreme weather event in this region, our action plan would be to avoid buying this currency. Even in the case of another cocoa-producing region experiencing an extreme weather event, we would still avoid buying the CFA franc currency despite the fact that the Cote d'Ivoire does not suffer from the dilution of exporting many types of commodities. In this particular case, we would avoid it because of the lack of our two key prerequisites, including political stability and liquidity.

So, to generalize, we find ourselves in somewhat of an investing predicament. We generally have found two classes of commodity currencies:

- **Class 1 commodity currencies** (examples: Australia dollar, Canadian dollar, Brazilian real, Chilean peso): These are the commodity currencies that meet our fundamental investing prerequisites of being in a politically stable region and having adequate liquidity in the foreign exchange markets (i.e., easily bought and sold in the foreign exchange market). However, they suffer from commodity dilution. In other words, the particular country in question exports a diverse variety of commodities and therefore will get only a marginal boost to the value of its currency in the event of a global climate shock in some other region of the world. It is, in fact, an indirect and second-string-type investment. *Therefore, our action plan for all Class 1 commodity currencies is to simply avoid buying them when they are the location of an extreme weather–based climate shock.* It would make a lot more sense to instead locate the direct and first-string-type investments discussed extensively throughout this book. Specifically, locate the appropriate action plan in this book for the particular commodity that is seeing the global climate shock.
- **Class 2 commodity currencies** (examples: many African-based currencies where only a single dominant commodity product is exported

out of the country such as with the CFA franc currency in Cote d'Ivoire, the global leader in cocoa bean production): This is the commodity currency type that satisfies the problem of commodity dilution because it has major export exposure to essentially one major commodity but at the same time fails because it lacks political stability and/or adequate liquidity in the foreign exchange market. For this reason, *despite having completely different characteristics than the Class 1 commodity currencies, the action plan remains the same. Specifically, the action plan with regard to Class 2-type commodity currencies is to avoid buying this currency in general but more specifically when it is experiencing a global climate shock of any kind. Instead, locate the appropriate action plan table in this book for the specific commodity type that is experiencing the global climate shock or extreme weather event.*

CHAPTER 13

Basic Principles of Futures Market Investing

As an investor in extreme weather events and in global climate shocks, there are ample opportunities to make money in many different industries and in many different financial markets. Occasionally, a particular extreme weather event does not lend itself to the more typical stock market–type investment. It may not even offer the opportunity for a corporate bond–type investment or an investment in the foreign exchange market. The only opportunity for investment, on occasion, is the futures market (or the equivalent exchange-traded fund [ETF] market). The purpose of this chapter is to remove the confusion and hence the fear associated with investing in the futures market and, as a result, open up an entirely new avenue of money making possibilities for you as an extreme weather–based investor.

We are going to start with a simplistic, high-level view of futures market investing without getting into any math or mechanics of operation yet. We simply want to talk about why an extreme weather–based investor would use the futures market.

As we have shown repeatedly throughout our global climate journey, extreme weather events often result in commodity supply shocks. A supply shock in any commodity means that the available supply of that commodity is severely reduced. The lack of available supply makes the price of that commodity go higher as a result. Because supply shocks generally represent a one-way street in terms of price reaction (i.e., up), this is primarily how we will make use of the futures market, to take advantage of rising commodity prices.

Without any additional complications, we simply buy a futures contract in a particular commodity if we believe the price of that commodity is going to go up. If the price of that commodity goes up as expected, then we make money. It is as simple as that. I mean, think about it—thankfully, these commodity futures contracts exist. If they did not exist, the only way to make money on rising copper prices would be to literally buy a giant pile of copper and then resell that pile of copper in the future after the price goes up. Doesn't it sound much more appealing to not actually take delivery of the metal itself but rather buy a future contract for copper and make money all on paper!

Now that we have covered the fundamental reason why and fundamentally how we enter the futures market, we are now ready to drill down a bit deeper into investing in the commodities futures market. Let's start out with a couple of important terms including the spot price and the futures price. If we were to go out into the open market to buy copper for immediate delivery the price quote we would receive is known as the spot price. By contrast we could be creative and buy a futures contract for copper on the relevant exchange and not actually ever have to take delivery of the metal. The futures contract is an agreement that mandates the holder to buy or sell the copper (in this example) at an agreed upon price and at an agreed upon time frame in the future. As the holder of the futures contract, if you so desired, you could actually take delivery of the metal at the expiration date of the contract, but in the vast majority of cases this does not happen. Instead, the contract is used as an investment vehicle whose value changes every day depending on how the current market spot price changes in a process known as the daily mark to market.

Our goal as extreme weather–based investors is for the spot price to continue to climb higher as a result of a global climate shock or extreme weather event, and therefore the value of our futures contract will get "marked to market" and also go higher as well.

You may be asking yourself whether the futures price is the same as the spot price, and if it is not, then why not? In the world of commodity futures markets, the futures price generally does not equal the spot price (time for just a little math). Other factors must be taken into consideration, including the cost of storing the commodity and the risk-free interest rate. More specifically, the futures price (F) is generically calculated from the following equation, where S is the spot price of the commodity, r is the risk-free interest rate, and T is the number of years or fraction of a year until maturity of the futures contract. We are initially assuming that there are no storage costs for the commodity.

$$\text{Futures Price} = \text{Spot Price} \times (1 + r)^T$$

A key to understanding what is happening here is that if the price of the metal does not change, then an investor makes zero profit on a futures contract. Another way of saying this is the only way an investor makes any profit on a futures contract is if the price of the metal changes. Therefore, at day one, when we buy a futures contract, the metal by definition has not changed in price because it is only day one. Therefore, the futures price can simply be thought of as taking the value of the metal in the spot market, S, and investing that value in a savings account with interest rate r for time T in order to capture the time value of money associated with letting the metal sit in storage for all of those months until the contract expires. The futures price theoretically has no other choice but to be this value. We can prove it by asking the question, what if the futures price were much, much higher than the value calculated in the preceding equation. If it were, then investors could make a profit even though the price of the metal has not changed, which is in violation of what we said in the first sentence of this paragraph. She could make this profit by borrowing money at interest rate r to buy the metal in the relatively cheap spot market today and simultaneously selling a very, very highly priced futures contract thus locking in what is called an arbitrage profit (i.e., a completely riskless profit). Even if this situation existed in reality with a futures contract price that was much, much too high, the investors who would be doing the profit trick we just talked about would cause the price of the futures contract to go down because there would be massive selling of that particular high-priced futures contract in order to lock in this arbitrage profit.

Now, back to storage costs. Recall from the equation that we temporarily ignored storage costs to make the point more clearly. Of course, storage costs must be added into the equation to calculate the price of the futures contract because warehousing metal, in fact, does have an associated cost. The equation does not really change very much and is still quite intuitive.

$$\text{Futures Price} = [\text{Spot Price} + \text{Storage Costs}] \times (1 + r)^T$$

The only additional comment on storage costs is that they are usually paid at the end of the storage cycle, and therefore the "storage cost" in the equation is actually referring to the present value of the storage costs.

The more mathematically inclined reader may be asking at this point, "What happens to the futures price as the spot price of gold, for example, changes? If the spot price of gold rises by \$100 per ounce, does the futures price rise by \$100 per ounce as well?" Using the preceding equation, assuming a risk-free rate, r, of 3 percent, with gold storage costs at \$50 per ounce over time period, T, of 6 months or 0.5 years, we can see the answer both in table format and graphically. Table 13.1 shows the relationship between the spot price and futures price of gold for given changes in the spot price of gold.

The current spot price of gold in the market at the time of this writing is \$1,545 per ounce. The most interesting finding in the table is located in

TABLE 13.1 Spot and Futures Price of Gold per Assumptions

Gold Spot	Gold Futures	Difference	Ratio
2045	2126	81	1.04
1945	2025	80	1.04
1845	1923	78	1.04
1745	1822	77	1.04
1645	1720	75	1.05
1545	1619	74	1.05
1445	1517	72	1.05
1345	1416	71	1.05
1245	1314	69	1.06
1145	1213	68	1.06
1045	1111	66	1.06
945	1010	65	1.07

Source: Theoretical calculation.

the "difference" column. The difference represents the futures price less the spot price. As shown at the current market price the calculated difference between the futures and the spot price is $74 per ounce using the above assumptions. Interestingly, as the spot price rises, the difference grows slightly, meaning the theoretical futures price on a dollar basis grows slightly faster. This table is very important because it gives the futures market investor comfort that when the spot price goes up, it indeed transfers directly to an improving value in your futures contract!

We can see the effect graphically as well in Figure 13.1.

As shown in Figure 13.1, the best-fit curve is a bone-straight line with the correlation perfectly positive. As shown from the equation and the

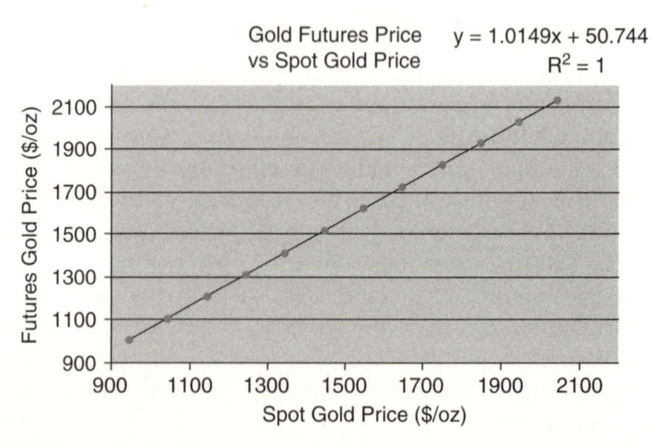

FIGURE 13.1 Spot versus Futures Gold Price
Source: Theoretical calculation.

given assumptions, the futures price will rise by $1.0149 dollars for every dollar rise in the spot price of gold. This should give the investor comfort because he knows that when his gold futures contract is marked to market he will be getting full daily credit in his contract when he sees the price of gold rise on any given day. Of course, however, this also means that if the price of gold declines on any given day, it will be marked to market on the negative side as well.

Let's move on to the high-level mechanics of how the futures market operates. For starters, I want to emphasize, as extreme weather–based investors, we will not be taking delivery of the commodity for which we buy a futures contract. For example, if we buy a copper futures contract for "delivery" in three months, we will close out our position before actual delivery takes place, similar to the way you would close out a stock position by simply selling the stock after you have made your profit. We simply desire to earn our profits on paper only. Interestingly, if we truly felt like it, we could, in fact, actually receive the full load of copper metal, which would be stored in a warehouse. Taking delivery of the actual metal under the terms of a futures contract is often inconvenient and in some instances quite expensive and in our case completely unnecessary.

A futures contract is referred to by its delivery month. The delivery months vary from contract to contract and are chosen by the exchange to meet the needs of the specific market participants. The exchange also specifies the last day on which trading can take place for a given contract. Trading generally ceases a few days before the last day on which delivery can be made.

If two investors agree to trade an asset at an agreed price at some point in time in the future, there is a risk that the transaction will not take place. One of the investors may regret the deal and try to back out, for example. One of the key roles of a commodities exchange is to organize trading so that contract defaults are avoided. This is where *margin* comes in, which we will cover next.

At the time you buy a futures contract for a particular commodity via your trading account, the broker will require the investor deposit funds into a margin account. The amount that must be entered into the margin account at the time the contract is entered is known as the *initial* margin, which will vary depending on the variability of the underlying commodity. The higher the variability, the higher the margin levels. At the end of each trading day, the margin account is adjusted to reflect the gains or losses to the investor in a process known as marking to market. The investor is entitled to withdraw any balance in the margin account in excess of the initial margin. In order to make sure that the balance in the margin account never becomes negative, a maintenance margin, which is somewhat lower than the initial margin, is set. If the balance of the margin account falls below the maintenance margin, the investor receives a *margin call.* He

TABLE 13.2 Futures Exchanges

Various U.S.-Based Futures Exchanges
CBOE Futures Exchange (CFE)
Chicago Mercantile Exchange (CME)
Chicago Climate Exchange (CCE)
ELX Futures (Electronic Liquidity Exchange)
ICE Futures
Kansas City Board of Trade (KCBT)
Minneapolis Grain Exchange (MGEX)
Nadex
NASDAQ OMX Futures Exchange
NY Mercantile Exchange (NYMEX)
COMEX
NYSE Liffe US
OneChicago, LLC

Sources: Public filings/data.

must then top up the margin account to the initial margin level the next day. If the investor refuses to top up the account, the broker will close out the position by selling the futures contract.

There are numerous futures exchanges around the world. The more popular of the exchanges located within the United States are shown in Table 13.2.

TABLE 13.3 Single Futures Contract Sizes

Commodity	Futures Contract Size
Corn	5,000 bushels
Aluminum	25,000 metric tons
Silver	5,000 troy ounces
Cotton	50,000 pounds
Cocoa	10 metric tons
Coffee	37,500 pounds
Oil	1,000 US barrels
Gold	100 troy ounces
Nickel	6 metric tons
Sugar	112,000 pounds
Copper	25,000 pounds
Wheat	5,000 bushels
Soybean	5,000 bushels
Natural gas	10,000 mm BTUs

Source: Public data.

These exchanges decide on the size of the individual futures contracts, depending on the needs of the most likely buyers. Because most buyers of these contracts are sizable entities, the size of the contracts is generally fairly large. A sampling of various commodities and the size of a single contract are shown in Table 13.3.

EXCHANGE-TRADED FUNDS

Exchange-traded funds (ETFs) are a remarkably simplistic and valuable tool. ETFs allow investors at all levels of experience to enter a vast array of financial markets. For example, the ETF ticker USO stands for the United States Oil Fund LP. This particular ETF gives an investor the opportunity to invest specifically in an asset that attempts to directly mirror the spot price of crude oil. Specifically, it attempts to mirror the spot price of West Texas Intermediate (WTI) light crude oil. Logistically, what is actually happening is that the fund is directly investing in futures contracts in the West Texas Intermediate Light commodity. This is enormously useful to the average retail investor because it provides the opportunity to enter numerous market types at quantities far lower than is typically found with larger investors entering the futures market directly.

ETFs are just another tool that we use to enter the various opportunities presented in the action plan tables throughout this book.

Final Thoughts

We have covered a lot of ground, literally, in terms of geography and geology. After covering all of the material in this book, you are now prepared with the three fundamental tools of being an extreme weather–based investor.

First, you have your financial market of choice. Whether you choose to stick with the stock market or to move into the bond or futures market, the investment ideas covered throughout this book can all be equally applied.

Second, you now have your list of the 17 rules of the extreme weather–based investor, as we discussed in Chapter 8, "Real-Life Examples: Execution, Results, and Timing," where we covered real-life examples of extreme weather and their impact on the financial markets.

Third, and most critical, you now have detailed action plan tables covering virtually any commodity in any geography in any extreme weather event around the world. You also have a detailed index and table of contents to help you rapidly locate the relevant action plan table once you get word of a global climate shock in any region of the globe.

The global climate condition that exists today provides ample investing opportunities for an extreme weather–based investor. As we have already covered, to the extent the global climate condition worsens, the more lucrative these investment opportunities become.

Whether you are a beginner or a Wall Street professional, you are now ready to begin your career as an extreme weather–based investor!

About the Author

Lawrence J. Oxley, an acclaimed author, initially received his degree in materials engineering in 1991 from the College of Engineering at Rutgers University, and later received an MBA from Penn State University in 2000 focusing on finance. In 2004 he became a CFA charter holder. He currently holds the position of Director, Global Research at the MetLife Insurance Company, with a focus on the Basic Materials sectors (i.e., chemicals, metals, mining, and forest products).

Index